Community Data

Community Data

Creative Approaches to Empowering People with Information

RAHUL BHARGAVA
Northeastern University

OXFORD
UNIVERSITY PRESS

Great Clarendon Street, Oxford, OX2 6DP,
United Kingdom

Oxford University Press is a department of the University of Oxford.
It furthers the University's objective of excellence in research, scholarship,
and education by publishing worldwide. Oxford is a registered trade mark of
Oxford University Press in the UK and in certain other countries

© Rahul Bhargava 2024

The moral rights of the author have been asserted

All rights reserved. No part of this publication may be reproduced, stored in
a retrieval system, or transmitted, in any form or by any means, without the
prior permission in writing of Oxford University Press, or as expressly permitted
by law, by licence or under terms agreed with the appropriate reprographics
rights organization. Enquiries concerning reproduction outside the scope of the
above should be sent to the Rights Department, Oxford University Press, at the
address above

You must not circulate this work in any other form
and you must impose this same condition on any acquirer

Published in the United States of America by Oxford University Press
198 Madison Avenue, New York, NY 10016, United States of America

British Library Cataloguing in Publication Data

Data available

Library of Congress Control Number: 2024939866

ISBN 9780198911630

DOI: 10.1093/oso/9780198911630.001.0001

Printed and bound by
CPI Group (UK) Ltd, Croydon, CR0 4YY

Links to third party websites are provided by Oxford in good faith and
for information only. Oxford disclaims any responsibility for the materials
contained in any third-party website referenced in this work.

To my wife Emily

who inspired and collaborated on so many ideas and projects in this book.

Contents

List of Figures	ix
Introduction: We Need a Larger Data Toolbox	1
How did I get here?	1
A new need	7
What to expect	12
1. Center Impact	16
A data protest in Mexico	16
A brief history of data	23
Libraries curate public knowledge	29
Journalists tell impactful stories	33
Museums reflect culture back to us	38
Democracy depends on participation	43
CSOs invite collective action	47
Own the impact of your data stories	49
2. Focus on Participation	51
A data mural in Philadelphia	51
Inspirations from pro-social settings	56
Critical theories of data use	61
Paint data murals	73
Build participatory data processes	85
3. Build Mirrors, not Windows	88
Racist data creation in the US	88
The concept of data literacy	98
Data sculptures as mirrors	109
Real world data sculptures	118
Use new media for data storytelling	129
4. Create Layers of Reading	131
A data fashion show in Tanzania	131
Problematic norms of data visualization	137
Act out your data	150
Build to complexity	165

viii CONTENTS

5. Open many Doors — 166
   A multimedia exhibit in Arizona — 166
   Hear your data — 172
   Smell your data — 180
   Taste your data — 187
   "Show up where they are" — 194

Conclusion: Practice Popular Data — 199
   Playful data in Los Angeles — 199
   Empower communities with information — 210

Acknowledgments — 216
References — 218
Index — 237

List of Figures

I.1. Data mural created by the Collaborate for Healthy Weight Coalition (August 2013).   2

I.2. A note from a participant, capturing the reflection that "we have never come together with data before, ever."   4

1.1. A mother carrying her remembrance.   19

1.2. Note the series of four dashes across the mid back of the central depiction of an ancient bull.   23

1.3. An ancient spreadsheet from the temple of Enlil at Nippur (1295 BCE).   25

1.4. The Map of the Future interactive climate change simulation.   39

1.5. One of the Neurath's health-focused exhibitions.   43

2.1. "*How We Fish*," a qualitative data mural in Philadelphia.   53

2.2. My simplified workflow for moving through a data storytelling project.   55

2.3. Participatory analysis of the GoBoston2030 questions dataset at an "idea roundtable" event.   59

2.4. Students interviewing lottery players in their neighborhood.   65

2.5. D4BL conference attendees writing on The Just Data Cube.   69

2.6. Levin's "Infoviz Graffiti."   74

2.7. Public visualization of average energy usage for residents on Tidy Street.   75

2.8. Errazurriz's tally-based mural comparing deaths among US Soldiers in 2009.   76

2.9. A team of volunteers working on a C4HW data mural.   78

2.10. The sketched out mural design collectively created with Kenyan graffiti artists Uhuru B, Bankslave, and Swift.   81

2.11. Offenhuber's Dust Zone (Staubmarke), 2018.   82

2.12. The second of three "Life Under Curfew" murals; residents used stencils to paint the responses on the mural itself.   83

2.13. The third "Life Under Curfew" mural involved tying strings to responses on a wall.   84

2.14. Resident Marshal Cooper paints a red line marking the historical zoning barriers.   86

3.1. The "average" woman from O'Brien and Shelton's dataset, as published by the United States Department of Agriculture (USDA) in 1941.   90

3.2. An auto rikshaw in India advertising data science education.   92

X LIST OF FIGURES

3.3. Closeup of the ostracon EA5634, recording reasons for being late to work. 95

3.4. Researchers speaking with community members in front of the REVEAL-IT! visualization of monthly energy expenditures in Barcelona. 96

3.5. A university community member moving their hand to vote in front of the MyPosition piece. 97

3.6. A screenshot of the summary of unidentified flying object (UFO) data presented as an example in the WTFCSV activity on the DataBasic.io website. 105

3.7. An Incan quipu. 112

3.8. A navigational start of the Marshall Islands. 112

3.9. Wooden intuit maps carved out of wood. 113

3.10. Columns of plants arranged to demonstrate the normal law of error. 115

3.11. Governor Cuomo in front of Mt. COVID-19. 119

3.12. 200,000 empty chairs set up for the October 2020 National COVID-19 Remembrance in Washington, DC. 121

3.13. Data from the Massachusetts Department of Transitional Assistance on SNAP enrollments in mid-2020. 123

3.14. The 1,659 sculpture on display at a local urban farm, where it was used to lead educational programing for youth about food access and food security. 124

3.15. A physical model of the Brooks slave ship, created in 1788 as part of the abolitionist movement. 126

3.16. The Domestic Data Streamers asked people to answer demographic questions by stringing thread between pins to create a parallel coordinate visualization, such as this 2014 version of their "data strings" installation. 128

4.1. A model walking down the runway at the Data Khanga fashion show. 132

4.2. The winning data khanga design showcases the shocking number of married women who have reported abuse from their husbands (50%). The *jina* (text) states "if you keep quiet, they will make you cry." 133

4.3. Bertin's visual variables. 142

4.4. Financial Times Visual Vocabulary Guide educational poster. 143

4.5. Mackinlay in 1986, depicting results from Cleveland and McGill's 1984 work. 145

4.6. Some of the zines created by youth at the Carnegie Library of Pittsburgh during the "Civic Data Zine Lab" led by Tess Wilson. 149

4.7. Students performing a data theatre piece based on data from Livable Streets Boston. 153

LIST OF FIGURES xi

4.8. Joiner's living histogram of students by height, showing a bi-modal shape due to inclusions of all genders. 158

5.1. Participants reacting to Ripple Effect. 167

5.2. A screenshot of a multiple-track composition in the TwoTone desktop application. 178

5.3. A visitor smelling the "smell map" of Milan. 185

5.4. Students baked brownies with varying levels of salt based on air pollution data. 189

5.5. Pizz'age by Léa Johnson and Alyse Yilmaz. 191

5.6. A scene from the CBS *This Morning* show segment on wealth inequality. 195

C.1. Los Angeles residents playing The Back 9. 202

C.2. A playable line chart from The Back 9 mini-golf course showing the narrowing housing gap. 203

C.3. A scene from the play being performed in front of an audience. 204

C.4. The "How to House 7,000 People in Skid Row" exhibit. 205

C.5. Visitors trying to "add it up" at the exhibit by moving boxes representing costs and housing units. 206

C.6. McLeroy et al.'s social-ecological model of health, as illustrated by Tasha Golden. 207

Introduction
We Need a Larger Data Toolbox

How did I get here?

In 2013 my friend Lisa Brukilacchio shared a data storytelling problem with me. Since you've picked up Community Data, it is one you've likely experienced as well. As a community engagement specialist in the Boston area, she had been working with local partners on a grant from the Collaborate for Healthy Weight Coalition (C4HW) initiative, collecting survey data about healthy eating habits and access to food in my hometown of Somerville. Data in hand, the team wanted to share it back to their community to collectively decide how to respond to their findings. Lisa knew a report full of bar charts and tables wouldn't meet this group where they were; it included immigrant families, local service providers, and managers of community organizations. Most of them didn't have experience with data analysis and the formal visuals typical to those kinds of reports.

That standard toolbox for sharing data offered a limited set of options to Lisa, and none felt quite right for her community, context, and goals. Charts, graphs, tables—none are designed to help you effectively engage communities in conversations, questioning, and decision making. These tools focus on understanding trends, demonstrating certainty, and pursuing truth and precision. The standard visual forms of data foreground scientific rigor, offering little support for developing engaging narratives or increasing community inclusion and participation. Science and evidence are critically important to communicate, but Lisa's community setting required a parallel approach. They were producing a more official report, but also needed something that aligned with goals of engagement and empowerment, something that would work for people who came from radically different backgrounds.

That same year my wife Emily and I were experimenting with new approaches to address that exact need for a broader and more appropriate toolbox for community data storytelling. We connected with Lisa and the C4HW coalition to create one of our first community *data murals*—a mural designed and painted based on collective analysis of a dataset by community members. The core idea was to learn from approaches that non-profits groups already

Community Data. Rahul Bhargava, Oxford University Press. © Rahul Bhargava (2024).
DOI: 10.1093/oso/9780198911630.003.0001

used to bring people together, like creating a community mural. We wondered: what would a mural created around data look like? In response, we planned a participatory process that moved the group through reviewing short data handouts, generating potential data stories and goals for telling them, collaboratively designing a visual image that tells that story, and painting it on a big wall.

We partnered with Head Start, an early childhood education program, and convened a group of parents, staff, and community partners to participate in the process we facilitated. Over a few design sessions the group narrowed in on a story they wanted to tell. Their key insight was that cultural context is critical for making healthy eating choices, and that local assistance programs need to recognize long held traditions of the home, culture, and community to help more effectively. Somerville is a diverse city of immigrants; the story they found in the data was built around their experiences of that cultural mix.

We painted the resulting design in the main lobby of the local Head Start enrollment facility (Figure I.1). The mural illustrated the data-driven story they developed, and incorporated data directly via symbols and text—flags of the survey respondents' home countries, city residents representing ages

Figure I.1 Data mural created by the Collaborate for Healthy Weight Coalition (August 2013).
Credit: author.

of participants, cartoons of real community members interviewed, talk bubbles with survey responses in different languages, clouds overhead labeled with highlighted challenges, buildings of locations shared as helpful resources (see "How to Make a Data Mural" in Chapter 2 for more details).

The mural beautifying their space was just one of the outcomes. The data story of challenges resonated with visitors to the lobby space, making them feel welcome, and validating their hardships. The mural named some of the issues they were facing in a relatable way. The data points about local services available told a story about how the community resources could help, connecting to local groups that already existed. Staff at Head Start walked away thinking about how they could produce and use data to improve their work, instead of how it could be used to monitor and assess them punitively. The key insights about cultural context and data analysis resonated into their work. Participants created connections and networks that still last to this day. Those interactions weren't secondary goals for Lisa and the team, they specifically wanted to build the networks that they suspected would help with the next steps of rolling out new interventions and programs. One participant shared a note he had taken in his notebook, which read "we have never come together with data before ever" (Figure I.2). This data mural brought together a disparate group of people, from a wide variety of backgrounds, with very different data literacy levels, to construct and tell an impactful data story that helped their local community address an issue of concern.

It is now common to bring people together around data in spaces that have little connection to the sciences or statistics. Like the C4HW coalition, these contexts require a broader toolbox with contextually appropriate approaches for using data, and new ways to represent information. Community organizations, museums, newspapers, libraries, local government—these are all parts of our society's social fabric where data use has quickly become commonly used. People in these sectors have always worked with information, but the forms and scale have shifted around and under them. The modern toolbox we have for working with and representing data wasn't built based on their needs and goals. This disconnect creates disengagement, disempowerment, and distrust. If you work with community data you need a more intentionally designed toolbox for using data, one that is aligned with your organization's goals.

With the C4HW coalition we used a variety of participatory data analysis and story-finding activities to meet people where they were in terms of data and visual literacy. The participants felt empowered to produce a mural that worked as a mirror back to the community itself. The goals were to create positive impacts on healthy habits, community cohesion, and individual efficacy.

4   INTRODUCTION: WE NEED A LARGER DATA TOOLBOX

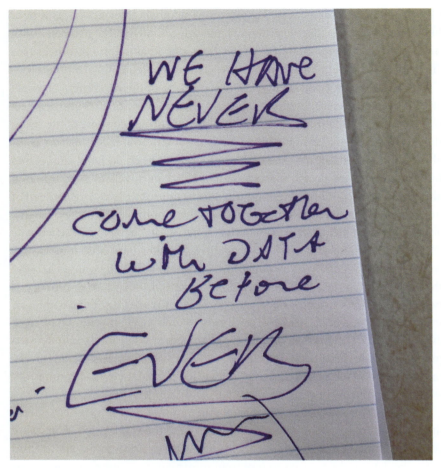

Figure I.2 A note from a participant, capturing the reflection that "we have never come together with data before, ever."
Credit: author.

Our facilitation techniques created another doorway into the data, offering playful and non-threatening pathways for finding a data story. The final mural is simple at first glance, but as you look at it longer, it offers more and more layers of data to the viewer.

> As an increasing number of public decisions are made around data, we need to find new ways to invite communities to have a seat at the table in the analysis and telling of stories with data, lest we actively disempower and shut them off from new data-driven processes.

Inspired by participatory and arts-based techniques, and driven by needs like Lisa's, *Community Data* introduces a framework for broadening your data toolbox to include approaches and media that are designed for community settings and goals. These techniques are already being explored by organizers, artists, journalists, activists, and creative data visualization designers across the world. This book shares learnings and examples created by those groups, connects to over a decade of my own work in community data literacy and data storytelling, and synthesizes a toolbox of alternative approaches to working with data and community. Putting this into practice will help you to use community data in the service of goals like empowerment, engagement, and justice.

A few years before working on the C4HW mural I was talking with my wife about a recent needs assessment survey she had done with local community groups outside of Boston. She works in public health, and at the time was assisting local groups that focused on issues like the prevention of addiction. The survey was intended to inform the types of training and support her organization would offer these community groups, and she shared with me that one of the top identified needs related to data and what to do with it. These community groups were being asked by their funders to report back data on their program operations and impacts, but they knew there was latent opportunity for action with the data. They wanted help understanding how data they produced could be leveraged to do their work better. I worked at a technology startup at the time, spending a lot of time analyzing data. Based on that experience and my background in education, we decided to put together a data storytelling workshop for this audience to see if I could help.

A few months later, in late 2007, I led my first Got Data workshop.[1] In the time since, I've facilitated over one hundred educational workshops for nonprofits, community groups, journalists, governments, librarians, museums, designers, and businesses. In 2010 I brought my nascent practice to the Massachusetts Institute of Technology (MIT) Center for Civic Media as a Research Scientist. I had built a rhythm of workshops and individual consultations with groups, to the point where one of my participants suggested a lovely new name for the effort based on how helpful she found it—and my Data Therapy project was born. The workshops continued and expanded internationally in 2014.

Over that time, I developed playful and engaging techniques to introduce these groups to working with data. My approach was built around the idea

6 INTRODUCTION: WE NEED A LARGER DATA TOOLBOX

that these new settings needed methods and materials that were more appropriate for their needs and goals. It began to form into a concrete set of activities that captured my pedagogy and style, focused on learning important skills in playful settings that were designed to appeal to people that didn't describe themselves as data people. This was based on my earlier time at MIT as a graduate student in the Lifelong Kindergarten group of the MIT Media Lab, where I learned the pedagogy of constructionism and an approach to learning built around peers, play, passion, and projects (as described by my advisor Mitch Resnick).[2]

In 2015, collaborating with Catherine D'Ignazio and supported by the Knight Foundation, I launched the DataBasic.io project to formalize and share the playful activities we had both been designing. Catherine and I first met when she was a master's degree candidate at MIT, where we collaborated on a number of techno-social projects. We convened a group of interested and related educators and hosted a series of collaborative design sessions to tweak some of the data literacy activities we had been developing together. I was using the activities in my teaching with MIT Media Lab graduate students, and she was using them in her teaching with journalism students at the Emerson College Engagement Lab. After each activity was piloted, we posted it on our DataBasic.io website. The website included tools with introductory videos, activity guides for educators, and technologies to support the activities, all in five languages.[3] DataBasic.io is still visited by hundreds of thousands of people each year.

Since then, Catherine and I have convened more collaborators to help us iteratively design and offer more than a dozen activities as part of the Data Culture Project, a lightweight self-service curriculum for learning to work with data in creative and arts-based ways. In 2018 we launched this program with support from the Stanford Center for Philanthropy and Civil Society and a learning community of thirty newsrooms, libraries, and nonprofits. The learnings I share in this book are built on experiences with the Data Culture Project, and related efforts that grew out of it.

As I interacted with more and more groups from different settings, I realized the growing need for an alternative set of inspirations for working with data and telling data stories. My partners had skills for working with information, they were just ones that didn't involve spreadsheets and charts. They understood the idea of posing questions to produce information that would inform decision making. However, the resulting data wasn't usually captured as large spreadsheets of numbers, and they certainly didn't call it data production and analysis. They relied on conversations, interviews, focus groups, observation, media capture, and other non-computational and non-statistical methods.

Partnerships and workshops work led me to question core assumptions about the processes and toolkits we use for creating impact with data. Why weren't we able to leverage existing information skills to introduce new techniques to these groups? How have historical uses of data informed our current practices? Why are visualization discussions so often limited to a specific small set of charts and graphs? Should we continue to privilege visual depictions of data over the other senses? Why do so many data literacy efforts seem focused on digital and statistical skill building? This body of questions has led me to create and highlight responses that look very different: data murals, data sculptures, data theatre, and beyond.

The method I now call *popular data* encapsulates those learnings and argues for creating more inclusive participatory approaches. The standard techniques of data visualization exclude large populations of people that participate in community processes. We can do better. Methods from the new settings of data use offer us an alternative path to bringing people together around data for the social good. I include this short summary of my own journey to offer you a sense of where I'm coming from and give you relevant context and motivation for the processes, examples, and related work I will share in later chapters.

A new need

It's been more than fifteen years since I led that first data workshop, and the world of data visualization is almost unrecognizable to what I introduced at that time. Hype around Big Data and artificial intelligence (AI) has continued to grow, but in response to some of the negative impacts being felt in social settings various movements have sprung up to offer pushback on this datafication of the world and our interactions. A first wave of popular press books has become new canonical critiques of data's use, including Noble's *Algorithms of Oppression*,[4] O'Neil's *Weapons of Math Destruction*,[5] and Eubanks' *Automating Inequality*.[6] These texts catalog ongoing real harms of data-driven algorithms in commercial and social settings and point to potential future impacts as ominous warnings. A second group of relevant books includes Benjamin's *Race After Technology*,[7] Criado-Perez's *Invisible Women*,[8] and D'Ignazio and Klein's *Data Feminism*,[9] all of which offered tools, processes, and theories to support a more reflective practice of data work, one grounded in the lived experience of people data often represents. A slightly

8 INTRODUCTION: WE NEED A LARGER DATA TOOLBOX

more recent third group of books pivots to focus specifically on data literacy and critical thinking in society, including Hall and Dávila's *Critical Visualization*,[10] Williams' *Data Action*,[11] and Loukissas' *All Data Are Local*.[12] The active and potential benefits and harms of data in society are being fleshed out, documented, and critiqued in books like those in these three groups. They've all influenced my thinking and practice.

In parallel to these books the open data movement grew up, turning arguments about government transparency into policies and laws across the globe. These efforts were built on the idea that data should be released freely by governments as a public good, available to drive innovation, increase economic growth, and support community work to better society. In the US this manifested in the OPEN Government Data Act of 2018, and similar policies were supported by the Organization for Economic Co-operation and Development (OECD) and the United Nations (UN) and their efforts across countries.[13]

The push to open data has created massive archives of information accessible to informed publics, about everything from weather, to budgets, to any other data related to roles the government plays. These datasets expanded on prior efforts, like the US launch of the data.gov index in 2009 as a central catalog of publicly available datasets. The open data movement hasn't been without its critics, nor its omissions, often driven by standard government partisanship and power struggles. That said, it has largely been a success, creating a civic expectation of data release and a legal structure in many countries to demand it.

Against this confusing backdrop of competing narratives, we're left to wonder—what is data good for anyway? Is it scaffolding for societal progress? Is it fuel for subjugation, marginalization, and discrimination? This book won't offer easy answers to this thread of questions but will argue that historical and modern practices are embedded in the processes and tools we use to work with data. When you work with data in community settings you must carefully and critically select the approaches you take to how it is produced, analyzed, represented, and used to support arguments. The dominant methods took their modern forms in the fields of science and statistics, but the massive growth in the use of data has brought those processes and media into new community settings, where the original assumptions they are based on do not necessarily hold true.

Data is hardly the objective truth many think it is. Humor me for a moment and imagine in your mind the archetypal family photo from a holiday gathering as a piece of data to discuss. How closely does it represent your family's reality? Is your cousin always as grouchy as she looks in that picture? Is your hair always so well put together? Do your two aunts really get along so well that they'd be hand-in-hand? And why isn't your favorite nephew in the photo at all? The family photo is just one depiction of the dataset that is your family; it includes some information, misrepresents some, and leaves some out.[14] That said, it sure is a wonderful memory and you'll cherish it in years to come.

The image in your mind of that imperfect family photo is helpful for thinking through the ways that data are encumbered with history. A dataset is partial, but usefully captures a set of observations. We can use it, but must do so carefully, and in concert with contextual knowledge that the data itself does not capture (whatever is outside the frame of the photo). This book, and the popular data framework I describe within, pushes you to constantly ask critical questions about how to work appropriately and effectively in the new community settings of data use.

The history of data is firmly connected to legacies of power and injustice. The production of information, and controlling access to it, has long been a tool to concentrate power and wealth. The census was created by the ancient Egyptians kingdoms to understand how much of their population they could force into labor on their pyramids. The British cataloged the slavery trade in monstrous spreadsheets still viewable in their archives. Wealthy individuals with early privileged access to insider financial information use it to accrue more wealth. I could go on (and will in subsequent chapters). How do we fight this history? How do we build a new practice of more contextualized, nuanced, and popularly accessible data? I argue that we need to look for inspiration in the new community-oriented organizations where data is being used.

We can do better than this history. Data scientists, visualization experts, and educators need to look far afield from their standard training. Arts-based collaborative practices, Indigenous data methods, participatory design approaches, multisensory museum experiences—these are the building blocks of a more just data practice for community settings. The activities in the Data Culture Project try to model some of these approaches, and I offer that as just one example for remixing, critique, and adoption by others as appropriate.

My approach is built on a particular pedagogy and set of design principles. In my time as a grad student at MIT I was introduced to the ideas of Brazilian educator Paulo Freire, whose concept of *educação popular* (popular education) focuses on inclusive, emancipatory, participatory, and egalitarian approaches to teaching and learning in the service of building critical consciousness. There's a lot to unpack in these theories and approaches (we'll get into it later). His ideas inspired me and radically changed how I conceived of learning. As you might have guessed, they led to the very name of the framework I describe. The central approaches and techniques (as well as the name) of my popular data framework incorporate key principles and lessons from Freire.

I create playgrounds for learning and building that serve very intentionally to complement formal software and statistical training. I've learned through iteration that hands-on, arts-based approaches to learning create new doorways for different types of learners, those often turned off by more traditional introductions to data. This work offers multiple ways of engaging with data and addresses gaps in the software-heavy data industry. The methods of popular data aren't about creating the most robust statistical analysis or the most evocative new interactive chart type; they are focused on practicing how to tell, construct, and convey a compelling data story targeted at a concrete set of community contexts and goals. We tell stories with data; a single dataset contains a multitude of potential narratives, serving as material for skilled explainers to work with. In an age of data being put in service of social good we need new translators for the public good. Practitioners of popular data serve this role.

Data is often discussed as an approach to finding answers. The practice of "solutionism" is rampant in data projects, offering people the idea that remedies to long standing problems can be found in de-contextualized innovation with technology.[15] The core myth of Big Data is that if you have enough data, answers will simply emerge. On the other hand, art is commonly described as a way to provoke better questions. Artistic pieces use a language of sensory experience to evoke emotional responses and reflection, inviting people into a space they haven't inhabited before, or to see something from a new perspective. If data is about answers, and art is about questions, how can the two work together? Both are necessary pieces of using data in support of action for the public good. If you're not asking the right questions, you can't suggest relevant answers. The language of the arts helps us to collaboratively find those key questions and offers a toolbox of techniques for engagement and participation. On the other hand, the world of data offers a rhetorical power of legitimacy and truth, powerful assets when fighting against the kinds of structural oppression that often impede social progress for the chronically marginalized in communities around the world.

How we use data matters more and more in society, and we're not using it in just ways. As data becomes more commonly used in social community settings, influencing the very fabric of our society, we need to focus more on how the methods and media of our data use connect to central goals of those spaces. Our use of data in pro-social settings must reinforce and resonate with goals of empowerment, engagement, and efficacy.

Throughout history those in power have marginalized groups by erecting barriers that prevented them from having a seat at the table for decision making in civic life. In many settings spoken language was the wall erected; to work with the ruling Roman empire in any setting they dominated, you had to speak Latin. The same held true for the British imperialists in India, where English was the key to entry. In modern times, written notices in multicultural communities often need to be translated by non-profit community groups to avoid leaving out their constituencies. In-person community meetings in the US are often held in the evenings, making it nearly impossible for those working night shifts, or parents juggling school pickups and dinners, to attend. Add to all this the recent pivot to data becoming more central in civic decision-making, and you have a whole new level of barriers to participation.

Depictions of data in public communication have become commonplace, they are viewed by large groups of people. This dramatically increases the responsibility for those working in data. AI expert Timnit Gebru described the history of this trend, noting that "data is the story of the victors, the story of the data collectors."[16] Uruguayan author Eduardo Galeano reminds us that "in Central America, the more wretched and desperate the people, the more the statistics smiled and laughed."[17] The history of data is fraught with injustice.

Data in these new sectors should bring people together in service of some community and social goals, lest we repeat this history. This is a very different approach from using data analysis and visualization as an objective input to a scientific decision-making process. To work on data for the social good in empowering and inclusive ways, we must embrace alternative approaches, pushing beyond the standard toolbox of extractive and visually centered practices we've gotten used to. We must find ways to authentically engage people who wouldn't describe themselves as math people.

The need is clear for me—a new path forward must truly welcome more populations into dialog with the ongoing systemic changes datafication has brought. We must meet people where they are, engaging their lived experience and knowledge. The core need is to help build a new set of approaches that push beyond just statistics, math, and technology, driven by goals of providing communities appropriate capacities so they can have a seat at the table in discussions of data processes that will impact them. These can exist in parallel to existing efforts, complementing them by creating new doorways for those that don't identify as spreadsheet power users or math whizzes, and don't strive

12 INTRODUCTION: WE NEED A LARGER DATA TOOLBOX

to become those. For participatory organizations that want to engage in data-driven processes as part of their work, we need new tools designed to bring people tougher to work with, and tell stories with data, in support of their goals. This contextually aware definition of the purpose of working with data is critical.

Data has moved from the lab to the library, from the boardroom to the community meeting room. We need a new set of methods and media to engage that change in setting. We must understand the limitations of the current media for representing data, and the core methods used in pro-social settings, to build this new toolbox. Popular data is built to respond to this need.

What to expect

In this book I walk you through my central arguments and approaches to working with data in support of community goals. You'll walk away with a new approach for your data work with and in communities, ready to challenge and critique the norms around you, primed to make collaborative impact with data, and inspired by novel and new examples. This is the goal and core contribution I hope to make with Community Data.

This book is built around concrete examples that will challenge what you think of as data storytelling, expand who you think can work with data, and introduce you to new settings where innovative data narratives are being created. I share many of my own experiences and work, but also interviewed a variety of practitioners whose stories you'll find relevant, illustrative, and inspirational. Rich photos of the work complement my writing to make it clear just how different this approach to data storytelling can be.

The chapters are organized around five key principles of popular data. Each begins with a real-world example, then proceeds to touch on related theory from academic research, practical examples from diverse fields of practice, and detailed case studies based on interviews.

In Chapter 1 I focus on the need to *center impact*. A data project created by artists, activists, and mothers of missing people in Mexico offers an example of a unique setting for data use. They turned memories and messages into physical data remembrances used in public demonstrations across the country. Digging into the history of our modern data approaches shows how they emerged from the realms of statistics and science, where the primary goals include evidence gathering, production of knowledge, and pursuit of truth. In contrast, I specifically focus on libraries, newspapers, museums, and civil society organizations where data is now more commonly being used to bring communities together. Examples from history and my own work in those settings highlight how their goals include empowerment, participation, efficacy, and education. These distinct goals often come into conflict and point to a key need to center impact when bringing people together around data.

Chapter 2 highlights the need to *focus on participation*. I summarize a visit to a mural in Philadelphia to look at how it helped the community collectively tell a data story. The *How We Fish* mural was built around collaborative data production to interrogate what work means to the community, resulting in an evocative mural that tells a story of independence, freedom and agency. Surveying four theories critical of our modern approaches to data, we find building blocks and complementary tactics that inform popular data—critical data literacy, data for good, data justice, and data feminism. Each offers useful lenses we can employ in service of rethinking mechanisms for working with data that are more appropriately designed for participation in community settings. I flesh out the process we use to create community data murals like the one for the C4HW coalition and connect it to other work putting data in public places to support civic deliberation and dialog. Data murals are a first novel approach for an expanded toolbox of data storytelling focused on community settings and pro-social organizations.

In Chapter 3 I look at how data is produced and used, arguing that we need to *build mirrors, not windows*. I open with the racist data production that led to women's clothing sizes being wildly inconsistent in the US. This underlies the datafication that has been built all around us, which uses data as a window to observe and make decisions about some other group of people. I review the movement to increase data literacy, which has been the dominant way of responding to datafication and how it connects to using data as a window. In contrast, the growing phenomenon of building data sculptures can help reposition data as a mirror, to be used by the subjects represented in the data to reflect on their own stories. This radically resets historical power dynamics embedded in the methods and tools of data, particularly aligned with needs and capacities of community settings. I review the work of academics, artists, and activists working on data sculptures, the materials we use to create them, and how communities interact with them. Data sculptures offer a second novel approach for our expanded toolbox of bringing people together around data in community settings.

Chapter 4 fights against the perceived complexity of data storytelling by proposing to *create layers of reading*. We start in Tanzania at a data fashion show, created by development specialists trying to put public data into use

14 INTRODUCTION: WE NEED A LARGER DATA TOOLBOX

and action. Textile designers built on local cultural uses of fabric for protest and speech to tell data stories in their designs about the scale and impacts of gender violence, and how to respond. From there I show how we can learn from interactive museum exhibits, where designers use a spark to intentionally design data narratives that pull people in, guiding them toward complexity and nuance in data stories slowly. I critique the narrow visual norms that dominate modern data visualization, arguing that they don't help us craft the layers of reading that we need to tell complex data stories to diverse audiences. As an alternative I summarize the long history of civic and deliberative theatre to craft theatrical productions that tell data stories and share results of my first experiments with others in creating data theatre for community settings. Data theatre is a third novel approach in the expanded toolbox, offering an alternative way to engage community audiences in data-centered discussions, creating a simpler alternate pathway into working with complex data.

In Chapter 5 I explore how we can use multi-sensory data experiences to *open many doors* into data stories in the new pro-social settings of data use. I begin by discussing *Ripple Effect*, a water-based exhibit designed as part of a project to democratize science around environmental data in Arizona. Complementing a data report about contamination levels, the multimodal exhibit used sound waves to bounce droplets of water above the surface to represent dangerous elements. Alternative epistemologies that push beyond scientific ways of thinking and representing data offer us inspirations for ways to connect with learners that don't take to computational and statistical forms of thinking. We must embrace multiple ways of knowing in our data stories to ensure we don't exclude those large portions of society. Creating data stories that work beyond the visual channel is one powerful way to find alternatives; we can hear, smell, and taste data. A growing set of designers, activists, and artists are creating sonifications, smellifications, and edibilizations of data, opening new sensory doors. These multi-sensory data experiences push the toolbox for telling data stories in community settings even further, offering new approaches and methods for engaging people in emotionally evocative and resonant experiences that draw them into topics they might not otherwise engage with.

Finally, the Conclusion synthesizes these arguments and examples with a call for you to rethink how you work with data in community settings, and who you choose to work with. To demonstrate this in action I summarize a set of data projects from a group fighting for housing rights in Los Angeles. They created a data-driven mini-golf course associated satirical theatrical production, and a hands-on interactive data exhibit about housing for the unhoused that exemplify my central principles. I close by summarizing the rationale and need for popular data.

You can change who gets to work with data, and how. The examples I share give you a new set of inspirations and methods for empowering people with information. My key principles lay out a roadmap for how. In our datafied society we must rethink how we work with data, and how we represent it, to ensure that a broader set of people have a seat at the table in data-centered processes.

1

Center Impact

The origins of data science are in the fields of statistics and science, where truth and evidence are key goals. The goals of museums, newspapers, libraries, civil society organizations, and civic governance are different. To work more appropriately and effectively with data in these settings we must learn from the methods and goals of how the pro-social sector works in community to promote the common good.

A data protest in Mexico

Buzz. Buzz.

In early 2023 artist and activist Martha Muñoz Aristizabal's phone was vibrating non-stop with notifications. Studies in the arts, data, and visualization hadn't prepared her for the emotional and logistical challenges of her latest participatory data project. Mother after mother from across Mexico were posting to share data about their missing loved ones. The women were spilling their memories into the WhatsApp group created to support her collaboration with documentary filmmakers, a project called *Te Nombré en el Silencio/I Named You in the Silence*. The film was telling the story of *Las Rastreadoras del Fuerte/The Trackers of Fuerte*, mothers searching for their missing ones in Los Mochis Sinaloa, in the north of Mexico.

Buzz. Buzz. "*Lo busca su mamá; No te voy a fallar, Te AMO*"/"Your mother is looking for you; I will not fail you, I LOVE YOU."[1]

Mother's Day for most is a time to spend with family, honoring the sacrifices and love of mothers individually and across society. Sadly, for many mothers in Mexico the day has evolved into an annual event more focused on sorrow than joy. In Mexico there are over 100,000 *desaparecidos*—sons, daughters, siblings, parents, and others who are missing, with no evidence of whether they are alive or dead. Every year on Mother's Day large groups of the missing's relatives take to the streets in protest to demand action, truth, justice, and accountability for the desaparecidos. A 2017 law required the state to create a database that would track them, but not until a court order in 2023 did the government announce

Community Data. Rahul Bhargava, Oxford University Press. © Rahul Bhargava (2024).
DOI: 10.1093/oso/9780198911630.003.0002

action from the National Forensic Data Bank to create a National Register of Unidentified and Unclaimed Deceased Persons.[2] In the meantime, this data hole has been filled by activists.

Buzz. Buzz. *"Papi estás en mi mente y mi corazón. Nunca dejaré de buscarte te amo."*/"Daddy you are in my mind and my heart. I will never stop looking for you, I love you."

Catherine D'Ignazio's *Counting Feminicide* documents how groups are using the practice of *counterdata* production across North and South America to document human rights violations.[3] Counterdata production names the work community groups perform to create datasets that document phenomena left uncounted by government and social institutions.[4] Formalized and informal coalitions are producing databases on missing people and using them to advocate for official investigations, protection, and accountability. In Mexico these advocacy efforts have included a series of artistic actions of remembrance, public spectacles built around the datasets of the missing that collectives have created.

Buzz. Buzz. *"Aunque quieran borrarte de mi mente cabes en la memoria de me vientre."*/"Even if they try to erase you from my mind you still fit in the memory of my belly."

The collaboration with filmmakers and activists that had taken over Muñoz Aristizabal's phone notifications was part of one of these public remembrances. The team was designing a data-centered spectacle of somber memories for performance across Mexico on Mother's Day in 2023. The filmmakers wanted to do a public action with the collective, based on all they had learned while filming there. Muñoz Aristizabal proposed the idea of creating spheres of remembrance, built to showcase the quantitative data the collective had produced and qualitative data including a photo and small letter from the family to their missing loved one. The spheres would be transparent, and embedded inside would be the photo and short message. The team created a WhatsApp group to collect this qualitative data, starting with just the mothers from El Fuerte who the documentary focused on.

Buzz. Buzz. *"La esperanza de abrazarte es el motor de mi existir."*/"The hope of hugging you is the motor of my existence."

The word spread quickly, facilitated by an invitation from the peace-focused Mexican non-profit Servicios de Asesoría para la Paz (SERAPAZ) to a video call including many mothers from other collectives. The team shared an invitation to the small WhatsApp group on the call and soon the group grew to twenty, thirty, fifty, one hundred, and beyond. This was participatory data production, yielding a spreadsheet of data about the missing

18   CENTER IMPACT

person, a photo, their name, and the message from their mother. The messages popping up in the notifications were powerful short notes to loved ones lost.

Buzz. Buzz. "*No pararemos hasta encontrarte mi niño hermoso*."/"We won't stop until we find you my beautiful son."

Speaking with me in an interview, Muñoz Aristizabal shared that the data "came from the ground, the people that were missing." The prompt to send a message to the missing centered the impact of their loss, and the potential impact of creating action based on the data. The response was so immense that Muñoz Aristizabal had to limit the number of spheres they could offer to make. The team simply didn't have the capacity to honor the contribution of every submission of data.

Buzz. Buzz. "*El infinito es poco, al amor que siento por ti*."/"The infinite is nothing compared to the love I feel for you."

With the dataset in hand, the team planned demonstrations and performances to carry out across the country. Each of the fifty selected missing persons was represented by a large open bowl, filled to about a third with soil from their hometown. Within that rested a transparent hand-sized sphere with a photo and message inside of it. On Mother's Day that year the mothers joined in the march across Mexico City. Beginning at the one-hundred-year-old Monument to the Mother, the group of mothers carrying their bowls led a crowd to the "*Glorieta de las y los desaparecidos*."/"Roundabout of the disappeared" and then to the "*Monumento a la Independencia*."/"Angel of Independence." A week later they presented thirty of the remembrances at the botanical garden in Cuilacán, Sinaloa. In each location the mothers carried their memory of this person in externalized physical form via the sphere, soil, and large bowl (Figure 1.1). The artifacts' design evokes metaphors of the womb, motherhood, birth, and the homeland. The powerful combination of physical elements, both symbolic and concrete, come together with the performative elements to create a grand spectacle of data protest. Muñoz Aristizabal described how people could "see themselves in this piece of art," holding up a mirror to the loss in the communities.

Buzz. Buzz. "*No descansaré hasta encontrarte*."/"I won't sleep until I find you."

The public events created a new way of representing this dataset. Muñoz Aristizabal had found that the few steps the authorities had taken to create official data about this crisis were "just trying to integrate information," while throughout the process the team on this project noted how "for us it is important to take care of the victims all the time." The experience of participating

Figure 1.1 A mother carrying her remembrance.
Credit: Amaresh V. Narro.

in, and watching, the performance with the spheres of remembrance created a moving impact beyond just hearing large numbers of disappeared people. They were using data art to impact the political system. For Muñoz Aristizabal this is critical, the art gave participants a "cathartic feeling" that they could be "integrated into the system of justice." The art gave them a chance to feel recognized, honored, and use their own data to advocate in a way that was "symbolic and emotional." This reinforces findings from science and environmental communications that point to the role visual art can play in driving participants and audiences to action.[5]

Buzz. Buzz.

Her phone is still buzzing today. The mothers continue to seek justice for their missing loved ones. The data-driven remembrances they are creating continue to make an impact. *Te Nombré en el Silencio* doesn't look like a typical data visualization. First, it exists out in the world, bringing data off the page and into real-life communities that the data is about. The raw data and demographic aggregations aren't left to wilt in some PDF report that few will ever read. Second, the demonstration hits you harder than a chart. When I first saw a video of the march in Mexico City it brought tears to my eyes, many of the quotes still do. The project presents the impacts of the loss the data represents in real human form. It speaks with human emotion in a way that provokes both memorializing and action. Charts speak with a visual

20 CENTER IMPACT

language of evidence and perceived objectivity that doesn't operate in the same way. Third, Muñoz Aristizabal built the whole project around those most impacted by the phenomenon the data represented. From sourcing the photos and messages from the mothers, to using soil from their communities, to having them hold the bowls and march—the community participated in every step. Fourth, the data performance was built around creating real-world outcomes to address the underlying harms the data cataloged. Charts are typically used to inform decision-making processes, combined as elements of a narrative that tries to make an argument or tell a story. The march and spectacle the mothers created was an argument, demanding action through public demonstration. These mothers were using the qualitative and quantitative data they produced to fight for power and voice. *Te Nombré en el Silencio* offers a radically different take on working with data and telling data stories, one that centers the community impact and draws on participatory and public processes common to activism and protest. Muñoz Aristizabal and other artists in Mexico are building an inspirational practice that produces impactful data experiences like this.

Learning from pro-social spaces

Over the last decades data has become common to see in sectors that seldom used it before. I'm focused on community settings, and accordingly center community-oriented institutions in this book. What do I mean by *community*? The word means so much to so many, so a concrete definition is hard to nail down. In general, when I use the word community in this book I am referring to a group of people that are invested in the relationships that bind them together, working together to improve their overall social condition. The communities that created, or are the audiences, for each case study look a little different. And the community-oriented institutions that interact with them? I'm specifically focused on libraries, museums, newsrooms, civil-society organizations, and government. For convenience I refer to these as *pro-social* organizations and institutions throughout. I evoke this name not to argue that science, statistics, and business are working against the common good, but rather to highlight that these pro-social spaces of new data use are centrally focused on those positive outcomes. The betterment of society is their main goal.

The first principle of popular data is that we must *center impact* in community data settings, like these pro-social groups do. Here I focus on centering positive impact, not negative. Since the mid-2000s, I've collaborated with numerous groups from these sectors. These pro-social organizations all work with information already, and I've seen many of the challenges they face, and solutions they generate, when they run into challenges with using our dominant forms of working with quantitative data. Their work is critically important to creating a larger toolbox for working with data in community settings.

Patrons are visiting libraries with questions about data access and use, but without existing data skillsets. In response libraries are developing creative hands-on public programming that fits the context and needs of those visitors.

Journalists have long produced and used information to inform and educate the public, hold power to account, and more. With an increasing amount of public knowledge being produced as large sets of quantitative data, the field has leveraged their information skills to innovate new ways of telling stories through the approach of data journalism.

Museums translate dense information-heavy topics into engaging experiences that reflect culture and science back to the public. With emerging concepts more often understood via large sets of quantitative data, museums have adapted their broad set of explanatory techniques to showcase and highlight data-centered exhibits that speak to visitors in novel ways.

Healthy democracies thrive on pathways for informed publics to engage in advocacy, voting, and more. With a growing number of civic planning and monitoring efforts built on producing and using quantitative data, our governance structures need to find ways to engage those that don't speak the formal language of data, lest they be left on the sidelines of data-centered processes.

Civil society organizations (CSOs) (such as non-profits) assist the marginalized by being able to open doorways to the rooms where powerful decisions are made, so their concerns can be heard and addressed. Quantitative data analysis has become that new norm in those settings, but many lack the training to engage around it. CSOs are adapting by translating that language for their constituents and helping themselves learn to speak it.

A short list of real-world examples can help flesh out how these settings aren't what our modern data practices were designed for. A business manager produces data about clocking in and out of work to track employee performance. A psychologist designs a test that attempts to quantitatively measure intellect. A physician measures structural dimensions of skeletons to categorize health outcomes. An analyst pores over data on purchasing to understand trends and potential for marketing. These are the types of settings our methods have emerged from, and many of them center around questions of power, control, and access. That business manager was using data to penalize employees for being late, creating a hierarchy of power that made employees suspicious of any data-centered programs. Both the psychologist who created the IQ test, and that physician who measured skeletons, were using the authoritative veneer of data to justify their racist eugenic beliefs. That analyst of purchasing trends was using data to create micro profiles of vast sets of individuals to market to them more effectively, and even pre-dictively. These examples are just a few plucked from the history of data being used to impact the subjects represented by data in potentially negative ways.

The new contexts of community data use look different. A government pro-duces population level data to understand and support societal shifts. A civil society organization produces data about resource delivery in small towns to hold the government accountable for their commitments. A journalist pub-lishes data they created about human rights violations to drive public attention to prevent them. A librarian helps entrepreneurial patrons find data to help understand the local competition their new business might face. A group of Mexican mothers produce their own data to protest for assistance in finding their missing children.

The contrasts of those two lists aren't manufactured; each is a real example (many described later in this book). Science labs aren't libraries. Boardrooms aren't community meeting rooms. Yet the methods and media we use for this data work haven't adapted sufficiently to these new settings to match their needs and constraints. This has created a disconnect that urgently needs to be addressed, because it is contributing to real harms and disenfranchisement being experienced right now.

Broadly speaking, a central approach of the pro-social sector is to help their communities engage with information in informal ways. This is a critical com-monality amongst the spaces I include under the umbrella of the term because informal learning spaces focus on impact in context. Information librarians Bowler, Acker, and Chi stated it concisely in a 2019 paper for the American Library Association:

> A key strength of informal learning, however, is that its primary goal is meaningfulness—learning that takes into account the sociocultural context of the learner and that is driven by the interests and motivations of the learner.[6]

Informal learning approaches in data projects open doors for more audiences by centering impact that learners can connect to. This stands in stark contrast to the history of data production and use, and the professionalized approaches of introduction based on computational and statistical methods. Answering the questions of where data comes from, and what we mean when we use the word, provides necessary context for understanding just how different projects like *Te Nombré en el Silencio* are.

A brief history of data

Humans have been creating data for as far back as our historical records go. The earliest evidence we have of humans making some kind of externalized representations of information are found in cave paintings. The ancient paintings of animals on rock walls have held a mystery that archeologists struggled to decipher for decades: many of them are adorned with oddly similar patterns of dots, lines, and Y symbols. These symbols are especially frequent in a series of caves in southwestern France that were painted 35,000 years ago (Figure 1.2).

Figure 1.2 Note the series of four dashes across the mid back of the central depiction of an ancient bull.

Credit: JoJan, CC BY 4.0, via Wikimedia Commons.

24   CENTER IMPACT

Amateur archeologist Ben Bacon was drawn to these curious symbols, which appeared across caves throughout the region.[7] They're a bit of a historical oddity, defying explanation since they were first discovered. Bacon had an idea, one inspired by their placement on animals. He surmised that perhaps they were some kind of data about the animals being recorded, serving as a form of proto-written counting system. Collaborating with a team of university experts, he fleshed out a theory that the marks served as tallies to indicate when the animals depicted were mating and giving birth.[8] Their explanation is that our cave dwelling ancestors were creating a dataset of animal fertility to track the reproductive cycles of populations they were hunting. The team proposed that the Y symbol, found close to the dashes in over 600 caves, indicated a birth. Our current understanding is that this was an encoded data system put into use more than 20,000 years before written language was invented. Survival drove the effort to record tallies of population and reproduction on cave walls.

We've been creating datasets for a long time.

Fast-forward to the time of the development of written language and we can find another relevant historical record. Ancient Middle Eastern societies offer the first evidence of people using data tables; our row and column-based format for recording data into tables has been in use for thousands of years. Curious about the population of lambs, goats, and sheep in the area around 4000 years ago? Clay tablets now housed in museums across the world can answer your questions. They left us the first evidence of someone recording quantities: rows and columns scratched into clay tablets.

Our clearest examples from the time come from the Puzrish-Dagan, the livestock management area of the kingdom of Ur 2028 BCE.[9] Their initial tables used whitespace and lines on clay to track ownership of sheep and goats in a grid of information—rows for each animal type, columns for each person, and cells for quantities owned.[10] About 700 years later we see even more complex tables being developed. Payroll tracking at the temple of Enlil at Nippur includes monthly payments to employees, and even includes subtotals for every 6 months (Figure 1.3). This is a 3000-year-old Excel payroll spreadsheet rendered in clay.

Bureaucratic administration needs led to the creation of systemic tabular recording systems. Based on the evidence from ancient times, we can see how counting and recording data was a driver of the development of written language itself. Those in power within a hierarchically organized society were using data to centralize and systemize their systems of governance; they needed to invent data tables to track and administer their kingdoms.

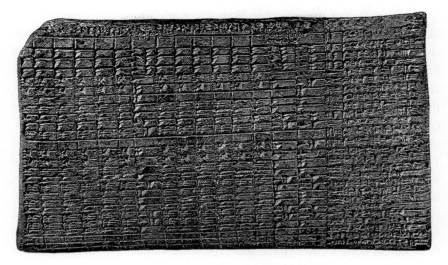

Figure 1.3 An ancient spreadsheet from the temple of Enlil at Nippur (1295 BCE).

Credit: A. T. Clay, Documents from the temple archives of Nippur dated in the reigns of the Cassite rulers. 1906, PL. V.

Our use of data and tables continued since then, from Ptolemy's astronomical lookup tables around 100 CE, to al-Khwarizmi's of 800 CE, to those of Renaissance era astronomers such as Kepler and Newton.[11] The 1600s saw massive production of tables for mathematics, such as logarithmic tables. These were one of Babbage's inspirations for developing the mechanical computer. In a very real sense, the data table connects directly to the invention of the digital computer technologies that have so fundamentally defined our modern times. The production and use of data is not a new phenomenon, it has driven fundamental developments throughout human history.

Defining data

What do you think of when you hear the word "data?" Most of us believe we have a shared understanding of what that means, usually related to collections of numbers. Perhaps you think of digital versions of the ancient Mesopotamian tablets. However, the history of the term is more nuanced. The usage and implication have changed over time. The etymology and evolution provide valuable context for how we think about data, who is allowed to use it, and what data represents.

26 CENTER IMPACT

Historian Daniel Rosenberg is likely the world's foremost expert on the history of the word itself, having published multiple academic papers and book chapters on conceptions of information over time.[12] Per Rosenberg's work, we find an original root in the Latin verb *dare*, which means "to give." In Ancient Greek mathematical texts and medieval religious writings data are givens, things that are known (vs. unknown). In the mid-1600s the terms became more commonly used in other domains, and more closely related to our use of "data" now.

In workshops I often ask participants to tell me what words come to mind when they hear the word data; most common responses include numbers, information, statistics, charts, and surveillance. The modern popular conception of data relates almost entirely to quantitative recording of information in formal ways; today "data" are numbers or spreadsheets to most. The rhetorical functions of these forms of data, and depictions, all draw power from the last few hundred years of the term's use—if data is thought of as a mathematical truth, then the spreadsheet is trustworthy, and the chart should be believed. This train of associative power underlies the rhetorical functioning of every data modern visualization. We're in a time now where too often data is referred to as truth itself, with no driving question guiding its use.

In mid late 2017 I found myself engaged in an illuminating discussion about the term with a Spanish speaker just outside of Rome. I had been invited there to run a set of workshops at the headquarters of the United Nations (UN) World Food Programme, whose data team was embarking on a large effort to position data more centrally in their decision making across all their regional offices. I was introducing the idea that often the simple act of counting is one of the first steps of data production, using the word *contar* with the mostly Spanish speaking workshop participants. I intended this to mean "to count," but my new Colombian friend pointed out how it also meant "to tell." In English the closest analog is perhaps "account," where "to account" for something could be counting, while "an account" is often a story of something that happened. This powerful duality in the terms connects the act of recording information to the act of telling a story; the same word is used for both.

The dual meaning of *contar* is critically important when we come to thinking about data and community stories. Over the last decades, we've been introduced to a diversity of processes offering to help move from data to story. Design firms, visualization experts, journalists, tech—they've all offered visual depictions of the way they model moving from data to story. Yet this definitional duality I'm highlighting suggests an alternate understanding, specifically that the initial act of counting itself is telling ourselves a story. This stands in contrast to the common popular conceptions, which link raw quantitative data (expressed in numbers) with truth and objectivity. Communities I work with want to tell stories with data, but they don't describe the data itself as a story. This lexical example helps me internalize, and explain, how impactful the ways we think about data are.

Fields of origin

The modern professionalized methods of using data and representing it emerged from math, science, statistics, and closely related fields. Fields such as economics, epidemiology, and cartography have also contributed greatly to our current approaches. From these domains of origin developed a shared set of principles that continue to underlie the modern preferred methods and media of data use. Math is a field of pattern finding, devising rules and systems to support logic-based deductions; it is in pursuit of demonstrably accurate theorems based on numerical operations. Science focuses on observation, experimentation, and evaluation of theories to explain phenomena in the world around us. The modern Western conceptualization of science is rooted in the idea of pursuing some truth or fact. The field of statistics, closely linked by most to data work, in part emerged from social scientists trying to understand the phenomenon around them.[13] Many connect it to probability studies and demography in the seventeenth century, where state actors were developing statistical methods to support demographic activities.[14] This was data used in support of centralized planning by those in power for governance (much like the ancient Mesopotamian kingdoms that produced our first data tables on clay tablets). Economics, epidemiology, and other numerically focused fields have similar histories. They all share an approach to producing evidence, making models, and assessing theories in support of understanding.

The fields of knowledge most closely linked to data have a history built around specialization and expertise. Each of these domains requires deep training to earn certifications and accreditation. They are not broadly focused on social good, although they frequently interact with it. They have rigorous methods and standards for common processes and have largely centralized around singular accepted ways of knowing; there is one standard shared language for doing math, a single driving scientific method.

Their approaches spilled over into the emerging field of data science in the mid-twentieth century. We don't know where the term "data science"

28 CENTER IMPACT

first emerged from, but various histories trace it to statisticians John Tukey, Bill Cleveland, and computer scientist Peter Naur.[15] They used the phrase to describe both the idea of applied statistics and the field that became known as computer science. Many attribute our modern popular use and understanding to data visualization pioneer Cleveland, who argued that applied statistics needed its own field, one that could be called data science.[16]

Over the last sixty years, from those pioneers, a consensus approach of how to work with data emerged. This has brought the methods, assumptions, and goals of working with data from those fields of origin to new domains of work. The sciences are defined by a pursuit of understanding and explanation, a drive to theorize the truth of something. More data in support of an operational theory makes it stronger. Look to our history of the mid 2010s and you'll see the evolution of this leading into an era of Big Data.

The driving idea of Big Data is that with enough data points the truth of something will just emerge, without even requiring the hypothesizing typically associated with the sciences. Like many technology terms of our time, the exact definition of *Big Data* eludes us (think of the "cloud" or "artificial intelligence" today as other examples from the world of technology). A seminal 2015 Boyd and Crawford paper provides the most relevant definition of Big Data for this discussion; pulling it apart into components of technology (computation and algorithms), analysis (pattern finding to support claims), and mythology (promising previously impossible objective insights).[17]

I extend this by pointing to four patterns of Big Data that connect directly to the fields of practice it emerged from and was informed by.[18] These offer a glance at how this history plays out in concerns and practices from real world settings of Big Data use; it is a set of tools employed by those in positions of power (defined here as the ability to exercise control or influence on other groups).

Big Data is opaque. Data about our interactions with the world is most often collected with only token approval from us. The instruments of data production and storage are disclosed to us in small print, and we agree with them as a routine practice with little understanding of potential current and future use. This denies us awareness of what data is being recorded and when.

Big Data is extractive. Those working with data extract it from us. It is most often collected by third parties, not meant for digestion, or use by us (the people actually represented in the data). This denies us any sort of processes for interacting with, or controlling access to, data about ourselves.

Big Data is technologically complex. Massive storage repositories and advanced algorithmic techniques present significant barriers to us using large data systems, even if we understand the technical jargon. This denies us any sort of understanding of how results of data analysis were created, and thus how we can critique them.

Big Data is centralized. Data analysis is usually handed off to an expert from outside our communities, and the results are used to make decisions that impact us. This denies us opportunities to participate in the decisions that have major consequences on our lives.

Like other methods of using data, Big Data embeds the historical practices of the sciences and mathematics that it emerged from. These methods and approaches didn't stem from practices in the pro-social contexts of data use, and in many ways are not well suited for them. Now applied in social domains that have real word impacts, this is leading to significant harms experienced by all sorts of populations.

The history of data and society reveals the connections that have driven our modern conceptualization of data, and the practice of data visualization. A singular focus on scientific methods of thinking and representing the world underlies most of the data science. The specific focus on the visual leaves us with little opportunity to engage with the potential for impact via arts methods.

Like the Mexican data protest, the pro-social sector offers a multitude of alternate paths to wielding the power of data. They look starkly different from practices of Big Data, and even computer-generated charts and graphs. Data science and visualization in community settings has much to learn from these alternate approaches, and the broader set of tools they rely on. Embracing a broader set of techniques leads to a more appropriate framework for working with data in ways that center positive impact as the key point like the pro-social sector institutions do. Libraries, journalists, museums, civil society organizations, and democratic institutions focus on justice, empowerment, participation, and efficacy. Learning from how they work is the way to build a more appropriate toolbox for work with data and representing it to achieve those community goals.

Libraries curate public knowledge

Visitors to Aarhus, Denmark have a variety of sights to enjoy—local cultural museums, a central Gothic cathedral, a circular boardwalk that juts out over the ocean just south of town. If you visit, amongst all of these I implore you not to miss the Main Library in Dokk1, voted in 2016 the best library in the world.[19] It has its own tram stop, a slide for children (and adults) hidden inside of a multi-story bear sculpture, a robotic book sorter, a well-stocked makerspace, a giant robotic gong, a wide variety of books, and public workshops for patrons of all types. The last item on that list of attractions is what drew me there. At the invitation of library staff I hosted a workshop on creative data storytelling as part of their larger efforts to introduce ways to work with data to the local population in 2019.

The modern library is an institution devoted to helping people from diverse backgrounds access knowledge in multiple forms. People come to libraries with questions. Over the last few decades more and more of those questions relate to data, paralleling the growth of quantitative data generated by society. The visitors asking aren't data scientists or business intelligence experts, they are parents, small business owners, young learners, and others. This has required libraries to come up with new ways to help patrons access that datafied knowledge, new ways to create positive impacts on the visitors and support their efforts to understand and shape the world around them.

Walking into the Main Library space in Dokk1, one of their most prominent architectural features you see provided a relevant backdrop for my trip—an eight-meter-tall gong created by artist Kristine Roepstorff swings overhead.[20] During my visit it was periodically stuck, revealing that it was somehow computer controlled. After asking around I finally learned that the gong was played whenever a new baby was born at the nearby hospital. This robotic piece of musical art translated data about the frequency of childbirth into sounds at the library. You couldn't ignore it when it happened; the reverberations echoed through your soles. As a percussionist, parent, and creative data storyteller, the piece spoke to me on multiple levels. The gong is an example of a very simple *sonification*—representing data as sound—a growing form of creative data representation (see "Hear your data" in Chapter 5).

A few years before my visit to Aarhus, local librarian Jane Kunze had a vision for the role their library system could play in these larger civic transformation efforts she saw happening. Kunze set up programs to bring in creative digital technologies and learning approaches in service of the public good for residents and library visitors.[21] As part of these efforts, the Aarhus Public Library system decided to host a multi-day Next Library event, focused on engaging librarians and the public to envision and design how the library should support learning and information in the twenty-first century.[22] Understanding how to work and tell stories with data was part of their list of core public skills required for our future.

I offered a short workshop at the event, facilitating hands-on activities that showcased non-computational approaches to representing data. My session introduced participants to ways that art pieces like the gong can bring people together around data in community settings. I led a group of librarians and the public through a series of activities designing physical data sculptures and other creative arts-based activities that invited people into data in new ways (see "Data sculptures as a learning tool" in Chapter 3). This group didn't want spreadsheet training, they needed tools they could use to create impact in community library settings. They knew their patrons needed a larger toolbox of techniques for working with data, beyond charts and graphs or spreadsheets and tables. The Next Library event created a space for them to learn and explore. The library in Dokk1 had grown far beyond the historical task of collecting and hosting collections of books, yet these data workshops were still true to its mission.

A short history of libraries

The core mission of libraries is to collect, store, and share knowledge. They are focused on published materials, be that clay tablets, sheets of papyrus, hand-bound manuscripts, printed books, or electronic media. Many modern librarians study within the formal field of Library and Information Sciences, where they focus on information organization, retrieval, management, and ethics. This history and training make them well suited for curating data, archiving data, and introducing people to data as community information stewards.

Library activities are all manifestations of the underlying goals of the modern public version of the institutions, which is to serve their community's knowledge needs. In the US context the history of public libraries took off in the 1800s, pivoting to a literacy focus and then working on social inclusion efforts through the 1900s. Since the early 2000s libraries have taken on a growing role as community hubs with social spaces, co-working areas, free wi-fi Internet access, makerspaces, public entertainment, educational programming, and more. Libraries are so stitched into the fabric of their locations, and trusted, that they are Federal Emergency Management Agency (FEMA)-designated essential community organizations, playing a role in emergency preparedness and information dissemination.[23]

Near my home, we can look at the Boston Public Library (BPL) system's latest strategic plan to see how one large system considers the motivations for this shift to serving as a broader community resource.[24] The BPL offers "services that anticipate and respond to neighborhood interests," existing to "to serve

32 CENTER IMPACT

and sustain communities that foster discovery, reading, thinking, conversing, teaching, and learning." This is a set of core principles that are about community engagement, education, and knowledge. These are the goals of the library, not limited to the logistical tasks of creating access to information. Libraries are community spaces that serve the knowledge needs of local populations in support of social inclusion, growth, and development. Their goals are to provide access and support to their patrons, and their mechanisms for doing so are the physical spaces they occupy and the public programming that happens within them.

Libraries and data

When you describe data as a form of knowledge we produce, it becomes quite obvious to see that libraries are already purpose-built to help us organize, store, and share it. Accordingly, libraries across the globe have designed metadata standards, data storage technologies, data access services, and data literacy programming. The role of Data Librarian is an established position with its own guiding mission and texts.[25] In academic settings, libraries have created massive data repositories to assist researchers in archiving and dissemination efforts.[26] In public settings, they have launched data literacy and access programs like those in Aarhus.

Models of data literacy in libraries, as in a broader context, tend to focus on statistics and visualization, spreadsheets, and chart making.[27] This is helpful for certain sets of patrons, those predisposed to mathematical and formal ways of thinking and computational learning skills. However, many of the patrons coming into libraries don't come with those skills and aren't there to learn them. In response, libraries have branched out beyond running data workshops for professionals, creating novel approaches like data zine workshops with youth and other creative projects (see "We need more approaches" in Chapter 4). Studying results from these types of examples, libraries have shared advice such as focusing on more than just technology, using local and relevant data, and working with community partners, as key ways to create impact.[28] If the modern library is for everyone, and offers data services and training, then it should integrate locally impactful and creative ways of working with data to broaden the groups they can appeal to.

The Pittsburgh-based Civic Switchboard Project offers a helpful example of libraries trying to meet the needs of their patrons in relation to data and learning.[29] They envision libraries and their staff as "core data intermediaries" to "support more equitable access to information." Civic Switchboard leverages the idea of the library as a core community institution well-positioned to interface between data producers, data capacities, and data users within civic settings. Their goals include network building, cultivation of organizational and individual capacities, and welcoming public programs. In 2019 the project offered case studies and instructional materials for integration of civic data into library science educational projects. Leaders rebooted it in 2023 to offer regional training institutes and network building across the United States. Structural programming like this, supported by the federal government's Institute of Museum and Library Sciences, demonstrate how the larger field of librarianship is defining, embracing, and supporting the key roles libraries can play in community data.

I've learned from multiple collaborations with libraries on data literacy and storytelling. In my hometown of Boston, I've led workshops on data analysis for library patrons and teenagers. Supported by the MIT Public Library Innovation Exchange (PLIX), I taught charts and graphs to entrepreneurs, youth, and unhoused residents at a library in St. Paul, Minnesota. I introduced approaches to browsing quantitative datasets to librarians at Harvard University. With colleagues at Emerson University's Engagement Lab, I co-hosted a video-based virtual data literacy training program for a network of US-based librarians. I'm hardly the only data literacy educator in the world, so you can easily imagine the diversity of data-related offerings that are available from libraries across the globe, beyond the examples I've shared. Libraries have helped respond to the growing interest in using data in pro-social settings in a variety of ways, all informed by their goal of supporting community inclusion, learning, and cohesion. Their responses to diverse groups of people interested in using data offers grounded ways to think about creating impact for anyone working in community settings.

Journalists tell impactful stories

In 1892 Memphis, Tennessee was a violently unsafe place for formerly enslaved Americans. This era is known as the Reconstruction, when slavery had been officially abolished a generation earlier, but the wounds of the Civil War and legacy of enslavement still festered fresh in minds. A group of Black and white youth arguing while at play was all that was needed to spark atrocities that eventually led to one of our earliest and the most impactful examples of data journalism.

34 CENTER IMPACT

Journalist and civil rights advocate Ida B. Wells is not typically on the list of founders of our modern data era, but she should be. Born during the time of slavery in the US, Wells was brought up in a politically active home, often reading the local newspaper to her illiterate father.[30] After her parent's premature deaths, she became a teacher to support her siblings and began encountering systemic racism against African Americans head-on. Wells fought to access first-class cars on railways, and wrote to speak out against the terrible conditions of schools she taught in. One of her most enduring data projects started in March of 1892, and brings us back to Tennessee.

Spring of that year was tumultuous in the area. A large new bridge had just been built nearby across the grand Mississippi River, bringing hopes of economic investment and revitalization to the city.[31] The warm weather of spring had come early in March, but then the weekend of the 17th brought the largest ever recorded snowfall, a full foot and a half.[32] Public streetcars were paralyzed, mail was left undelivered, two freight trains collided and derailed, and the scheduled St Patrick's Day parade was canceled.

That spring also saw the violent lynching of Thomas Moss, the manager of a Black co-operative grocery store owner in town, and a friend of Wells.[33] A week earlier white and Black youth got into an argument outside his People's Grocery store, quickly growing into a large brawl between the two racial groups. A series of violent confrontations and police engagement eventually led to a group of white men dragging Moss and others to a local rail yard, where they were executed via beating, shooting, and lynching. All this was cataloged and reported by local white journalists. The incident is remembered in our history books as the People's Grocery Lynching.

Wells was a member of the grocery, a friend to Moss. She was inspired amidst her sorrow and rage to write publicly about the event, and the deeper pattern of killings of Black Americans, specifically focusing on the terrible spectacle of lynching. Wells began to tour the US, cataloging lynchings to build a dataset to show the sheer scale of their impacts.

For modern residents of the US, it is hard to imagine the role lynching played in social life across the South in the 1890s. They were used to strike fear into the hearts of Black civil rights campaigners, voters, merchants, and others. Many were planned ahead, inviting white viewers to bring picnics and later sold horrific postcards made to "commemorate" the events.[34] Thanks to Wells' copious and thorough journalistic research and data production work we now know the scale of that stain on American history. Her *The Red Record* book used reports in mainstream white newspapers to create a catalog of lynching and connected allegations, eventually finding more than 10,000 victims in the last thirty years of the 1800s.[35] She presented the report to President William McKinley in 1898 and went on to co-found the still-prominent National Association for the Advancement of Colored People (NAACP).

In journalism circles Wells is well known as an early innovator of what we now call *investigative journalism*—when reporters undertake deep and thorough information collection in support of a single topic. In data circles her story is less told, yet it is an important story of journalistic data production and analysis that centers impact. In *The Red Record* Wells admonishes us to "tell the world the facts and some means will be found to stop it." Her methods and tabulations were thorough, well-documented, and reproducible. She hoped that a society seeing the scale of the injustice laid bare in data would move to prevent it. Wells inspires us to retake our data and use it for our own purposes, fighting entrenched power structures that might deny us justice. Her work helps us question the goals of data production and use. Who is data for? How is it used in society? Who gets to produce it? She centered the impact of the data in her journalism work, trying to use it to prevent the systemic violence against her people. This is the core contribution journalism has to offer to broader sets of data storytellers, that we should center impact.

Journalism in society

I grew up in a household regularly tapped into global news via mass broadcast media but had little understanding of how much news consumption was shaping my perceptions and impressions of the world around me. The newspaper arrived every morning and we watched the national and local evening news show every night after dinner on TV. Sunday morning was frequently for national political talk shows, and any diaspora-hosted shows on local cable access. My parents immigrated to the US in the 1970s from India, and my family was scattered around the globe. Keeping up with global news was a way to keep in touch with events unfolding around them, wherever they lived.

I didn't necessarily pay attention to those information streams while running around in my youth, but the idea that keeping up with global news is important has stuck with me into adulthood. I still check in on multiple traditional news sources every day—from the British Broadcasting Company (BBC) online to my local National Public Radio (NPR) news radio station. That pre-digital era news exposure played a large part in taking me from studying electrical and computer engineering to becoming a professor in a School of Journalism. Journalism keeps us informed about events that shape the world

36 CENTER IMPACT

around us and helps us understand their impacts at multiple scales. Journalism shapes our perceptions of states, cultures, and individuals. Journalism pushes us to act on narratives that evoke our curiosity, empathy, rage, and hope.

There are various theories describing how journalism has those impacts. Political communication researchers McCombs and Shaw's formalization of *agenda setting* theory proposes that journalistic mass media dictates what issues are important to think about, even if it doesn't imply what to think about those same issues.[36] Philosopher and anthropologist Gregory Bateson's theory of *framing* digs into the latter, offering one take on how rhetorical and narrative tools are used in mass media to shape the opinions of news consumers.[37] Scholar Michael Schudson argues that journalism plays a core role in maintaining democratic society via informing the public, investigating power, explaining complex events, bridging between communities, offering fora for conversation, mobilizing groups, and showcasing the practice of democratic norms.[38] Labels for modern journalistic practices now include constructive journalism, solutions journalism, reparative journalism, and more; it is a field undergoing constant change and innovation, driven by more than just technical disruption.

The core goals of journalism revolve around serving the public good via sharing information in narrative form. Mass media is often described as the *fourth public estate*, complementing the legislative, executive, and judicial estates established to drive modern democracies. However, while serving the public good, news organizations often leave unstated exactly which publics they serve. During social movements of the US 1960s a non-mainstream press pushed back on dominant narratives from mass media and the perspectives they left out, describing their counter-culture work as the *fifth estate*. Some continue to use this term to define influential digital native news sources of the last two decades, further painting them in contrast to the entrenched mass media; I describe Wells' work in this way.

Data journalism

Journalism has been radically transformed due to digital information technologies. Local news, the most trusted news source in the US, has been gutted by the loss of advertising and takeovers by hedge funds.[39] Journalists are inventing new types of storytelling techniques using interactive digital tools that push beyond standard passively read narratives.[40] Yet producing and sharing civic information is still one of the driving roles local journalists can continue to play despite these challenges.[41] As more of our society and civic life is informed by quantitative data processes, media companies have responded with more data journalism.

This isn't a stretch; journalists are exceptionally well-positioned to work with quantitative datasets because of the field's history and training with information gathering and vetting. The core practices of modern journalism all translate to working with the quantitative forms of data that we more often encounter now. Interviewing is thoughtful and intentional data production. Fact checking shows a deep commitment to validating the sources and accuracy of the data being used for a story. Interviewing multiple sources on a story is a strong approach to assessing bias in the data produced. The diversity of narrative forms in journalism demonstrates a rich body of storytelling techniques to build on.

This isn't to say all is well and good in the history of data journalism. There is a well-established pattern of poor visual depictions of data in newspapers, most exhaustively cataloged in Murray Dick's *The Infographic*.[42] Visual distortions and omissions were regularly used throughout the growth of the infographic as a form in journalism during the early 1900s.

Luckily things have gotten better (i.e., more accurate) since then. There are many forms of data use in journalism, but all are connected to journalistic goals of constructing narrative experiences that weave in and around data. Ida B. Wells' spirit lives on in new forms of investigative journalism that are now called *open-source investigations*.[43] These include techniques for auditing and verifying social media videos of human rights atrocities,[44] producing important data sets of events such as police shootings of civilians that are otherwise untracked,[45] and building software to drive reporting on discriminatory algorithmically driven pricing models.[46]

The predisposition to use information, and the related skill sets journalists already have, has led to journalism being one of the primary places for innovation in data storytelling technologies. Many core data visualization and interactive website development platforms have emerged from or been heavily influenced by news organizations. For example,

- d3.js: the core author of this fundamental low-level webpage manipulation library, Mike Bostock, spent four years working at the *New York Times*.
- Flourish: the data visualization and interactive narrative tool for non-coders was co-founded by Duncan Clark, an experienced data journalist.

38 CENTER IMPACT

- Svelte: the fast-growing Javascript component framework was created for journalistic needs by Rich Harris while working at the *Guardian* but has found use far beyond the sector.

Many of these technologies are released as free and open-source software, allowing anyone to use them and join the community of coders that build them, offering yet another contribution to the common good.

Another interesting parallel is that, like data, journalism carries a rhetorical power of authority (if not accuracy) built on the history of its development. However, the young modern practice of journalism has more recently faced questions of legitimacy and relevance that parallel critiques of data. A key recent criticism is of the concept of objectivity as practiced in journalism since the mid-1950s. Led by Black journalists, such as Pulitzer-prize winner Wesley Lowery, there is a growing discussion about whether the term has in fact cemented a default white male perspective amongst editors of traditional news.[47] The results include stories that are deferential to police narratives instead of investigating what happened at an incident,[48] quote politicians from both American parties even when one is demonstrably lying,[49] and deny reporters agency to report on events related to their perceived identity-based community.[50] What is the alternative to this historical concept of objectivity? Journalism that highlights issues and stories from the marginalized, holds itself to account, and in general works to uphold democracy and freedom as core goals. Much like the dominant data processes and forms discussed in this book, data journalism is wrestling with a history of power and agency and how to be in the service of the social good.

Museums reflect culture back to us

Climate change is one of the key challenges of our time, across fields such as science, policymaking, international governance, and public communication. There has been significant academic work documenting the history of failures in climate change communication, and lessons they have to offer.[51] In 2016 I engaged that history head-on while working on an interactive museum exhibit, a table-sized computational climate change simulation called *The Map of the Future*. At the time I was working at Tactable, a maker of interactive tables that worked much like a giant iPad. We had been commissioned to create the piece as part of a National Science Foundation (NSF)-funded exhibit about the global and local impacts of climate change, which would then travel around the US for three years.

The design team decided to focus the experience on actions we could collectively choose to take right now, and the climate and social impacts those decisions would have in the decades that followed. Like most people at the time, I had not actually read the Intergovernmental Panel on Climate Change (IPCC) reports and was surprised to find they cataloged solutions and projected positive impacts. *The Map of the Future* was designed to represent the data in those reports via categories—forestry, wind energy, nuclear power, etc.—allowing museum visitors to turn the knob on each and see the impacts on the world of 2075.

The resulting piece was a technologically complex interactive data simulation (Figure 1.4). The table's surface projected an animated map controlled by physical dials that could be placed on it (cameras underneath tracked the dials). Each dial controlled an energy demand or supply. Turning a dial on the surface changed the resulting level of energy flowing into the model from that source, and the model's results were shown in a variety of ways in the animation. Under the hood was the internationally recognized Climate Rapid Overview and Decision Support (C-ROADS) climate model.[52] On the surface

Figure 1.4 The Map of the Future interactive climate change simulation.
Credit: author.

40   CENTER IMPACT

visitors saw a thermometer in the center indicating global temperature rise, floating dark clouds reflecting total carbon dioxide levels, and short videos showing the human experience of the changes. Reducing fossil fuel use might show a synthetic news report about energy riots breaking out. Turning down agricultural land use might pop up a video about how we're all eating insects.

This piece is still the most challenging interactive data visualization I've worked on, mostly because of the needs of the museum setting. Science museums invite in curious visitors with very informal rules of engagement; you need to design exhibits for interactions of five second to five minutes, for groups of one or twenty people, for facilitated and unguided experiences, and even more situations with diametrically opposed constraints.[53] One visitor might know nothing about the topic, another might already disagree with the premise, while a third might be a domain expert. You have very little control over the context. This is a crazy set of challenges, and it is incredibly hard to design one piece to satisfy all of them.

The Map of the Future piece succeeded in some ways and struggled in others. Interactive educational exhibits are often judged against the "all/most/some" criteria for assessing learning outcomes. I hoped that all visitors walked away with the idea that the climate is changing, most walked away understanding that human energy production and use is a key driver of climate change, and some walked away with context for how impactful each type of energy is on our human experience and the climate over the coming decades. Complex data stories are hard to tell, and museum settings forces you as a data storyteller to embrace multiple layers of reading and engagement in a way that is valuable to consider in other community contexts.

The museum as a gathering space

People go to museums looking for experiences, knowledge and information, and museums have a goal of making it accessible to the public. They used to be privileged places where owners held artifacts of knowledge and controlled access so only those with money and power could enter.[54] During the mid-1800s across Europe, many began to welcome the public with displays of tantalizing oddities. Following emerging world exhibitions and fairs, museums began to take on more of a defined role as locations for the dissemination of knowledge to the public. The public museums that grew from private collections were organized by rationalist taxonomies. They are centers of learning, but of course exist within the larger world of socially situated identities of the visitors. The modern museum has moved beyond that to be a public institution that researchers, collects, conserves, and shares culturally significant objects. In our current times art museums curate cultural heritage, natural history museums promote understanding of our environment, and science museums classify and inspire visitors with displays of the results of our curious questioning of the world. Museums support civic cohesion through reflection of collective identity and serve as cultural memory for both atrocities and achievements.

An idea of the educational museum emerged in the context of the growth of public schooling and questions of where public education should happen.[55] Science and children's museums grew from this, dealing with exhibits centered on information especially often. Now they offer interactive experiences that educate visitors of all ages; our *Map of the Future* is one example. Pedagogically they tend to serve as informal learning spaces, in contrast to settings of formal education like most schools. Part and parcel of this work is their role in translating contested topics and histories, sparking debate about the stories we tell ourselves about science and history. Visitors generally come with questions, and openness to learning about information-driven topics like climate change, but with vastly diverse backgrounds. Some groups certainly react to charts and graphs, but many others need different solutions, like an interactive table-based simulation, to find doorways into phenomena that are best understood through data.

Museum practices are sometimes described as part of efforts to create a *third space*—a destination where individuals can participate in co-creative practices that give them a sense of ownership like they feel in the home (first) and work (second) spaces they inhabit.[56] Many museums intentionally operate as civic spaces that stitch a community together around and in dialog with the artifacts they contain and display. This approach has been driven by several inspirational and impactful figures over the last 150 years. John Cotton Dana pushed for community engagement and holistic learning.[57] Stephen Weil argued for museums shifting from being about something to being for someone.[58] Frank Oppenheimer's Exploratorium in San Francisco offered new approaches to open inquiry and hands-on exploration.[59] Elaine Heumann Gurian put the museum in the context of society in a push for inclusion and social relevance.[60] Nina Simon advocated for museums as community-centered spaces for co-creation and conversation that can bridge across social divides.[61] All of these key approaches and conceptions connect deeply to the idea of educational museums as cornerstones of community, especially when it comes to data and public understanding.

Museums and data

Not immune to the reach of datafication, museums have increasingly acquired data visualizations, offered up datasets about their collections, and built data-centered interactive exhibits. The New York Museum of Modern Art in 2017 acquired Lupi and Posavec's *Dear Data*, a collection of hand-drawn data visualizations of their activities in daily life created over a year of cross-Atlantic correspondence between the two authors. The Tactical Technology Collective, a Berlin-based non-governmental organization (NGO) focused on digital civil society, created *The Glass Room* as a traveling exhibition exploring the dynamics of datafication and negative impacts on society.[62] Hosted at museums and other settings across the globe, it creates a space to engage public groups in reflecting on the negative impacts of datafication on themselves and others.

The development of Otto and Marie Neurath's *Isotype* offers another strong connection between modern data visualization and museums. Today's comic-like infographic style has deep roots in the Neurath's International System of Typographic Picture Education.[63] More critically for this discussion, it was birthed and first practiced under the auspices of designing a museum based around visuals of information that would be easily remembered (Figure 1.5). Neurath ran the Museum of Economy and Society in Vienna in the 1920s, built around visual depictions of labor, culture, and housing information designed to reach the less-educated working-class population. The concept was to create a more democratic museum, one that focused on showcasing modern life and processes in understandable ways, specifically focused on the visual. Their underlying belief was that access to wealth could be increased if knowledge about society was more easily available.[64] The Isotype system was designed for this type of "universal" communication, utilizing a standardized set of simple human figures and icons, laid out in hierarchical structures to depict flows, and emphasizing labels and text to focus on educational goals. The museum setting helped define Neurath's Isotype, a hugely influential visual grammar still in use today. If you've read or designed an icon-based infographic or flowchart, you owe some thanks to their work.[65]

Museums reflect ongoing conversations and trends in the broader world outside their doors, so it comes as no surprise that they are engaging with data in various ways as its production and application continues to transform our lived experiences and social fabric. Museums create interpretive experiences for publics that visit them, offering a language and space for explanation and engagement with data. They are a place that meets people where they are, in non-professionalized contexts. We can look at the rich set of formats they use

Figure 1.5 One of the Neurath's health-focused exhibitions.
Credit: Otto and Marie Neurath Isotype Collection, University of Reading.

to build a broader and more appropriate toolbox for bringing communities together around data, leveraging their power as cultural institutions.

Democracy depends on participation

In the mid-2000s I found myself wandering around a local square in my hometown of Somerville, just outside of Boston in the northeast US, looking for a community meeting I hoped to attend. I was excited because this gathering was going to focus on sharing data with the community; the mayor was launching a new program of data-driven updates about government functions. This local effort was riding a wave of global projects to have cities run more efficiently and effectively by producing and making decisions based on quantitative data. As an educator and researcher just starting to work on data literacy at the time this idea was right up my alley.

That evening I had a quick snack after work and headed over to the address on the announcement. After poking around lost for a bit I finally came across a small door that led down a set of poorly lit stairs. Making my way down carefully, I headed toward the low hum of voices I heard at the end of the

44 CENTER IMPACT

hall, eventually emerging into a small theatre with a central rectangular stage and seats arranged around three sides of it. There were a few dozen residents already there, all chatting or watching the city staff prepare the projector and laptop computer on the central stage.

Soon the staffer turned and began a run-down of city operations, all described with data in some way on the slides behind her—bar charts showing reduction in road pothole counts, maps showing locations of reports of rats, scatter plots of recent improvements in school standardized test scores. The last one gave me pause because I struggled to understand the visual. The presenters had put together an animated bubble chart that I puzzled over for a while, quickly losing track of the talking points being shared by the speaker. There were lines, circles, colors, multiple vertical axes ... I couldn't tell if they were showing a change or a rate of change, data for my city or multiple ones, just one year or a set of years.

I was lost.

This chart had me totally and utterly confused.

As I slowly looked around to see if others were getting it, their faces revealed that the complex visual depiction was confusing them as well. The speaker's narration wasn't helping clarify things for me. I think they were trying to tell a story about how our city's standardized test scores were still low but improving more quickly than others. Then again, I'm still not quite sure.

This visualization was far too information dense for the audience and setting, the story far too detailed to understand without significant background context. The staff's presentation of this data didn't meet this audience's visual literacy and capability for understanding data at all. We were being actively disengaged from the process by the presentation; the data visualization was making us less likely to participate even though the whole meeting had a goal of citizen participation.

Democracy and power

Democracy is a fragile but dominant ideal for governance in modern times. Representative democracy is characterized in detail by political scientists, philosophers, politicians, and others in a variety of ways. A few key shared tenets of democracy relate to my arguments.

- *Popular sovereignty* dictates that the power of self-determination lies with the people.

- *Political equality* requires that engagement in democratic processes should be inclusive and free from barriers.
- *Pluralism* is the idea that people of different backgrounds, beliefs, and opinions can all be embraced in a democratic society.

These three points don't define modern democracies, and also sound idealized and far from the realized practices so many of us live within. Our democracies are typically broken systems confusingly structured to both uphold and upend core ideas such as these. The mechanism and structures of central tenets are being directly attacked in the US and beyond by powerful forces. Many of these attacks are built on datafied structures created over the last three decades. Existing legal human rights frameworks are inadequate for the datafied and digital world; they were developed for a different time. Access to information has long been controlled to exert power; datafication of governance risks following in this history. How can we ensure data doesn't reduce sovereignty? Where does the government use data to support or impair equality? How can data be more pluralistic, capturing a wide range of perspectives on a topic? I don't claim to have easy answers to all these provocative questions, but I certainly engage them all via case studies and related theories.

The idea of the temporary creation of *mini-publics* is one way to realize principles like these in action, sometimes taking the form of civic meetings in public spaces which push governments to invite engagement outside of standard voting cycles (like the basement meeting I attended). Mini-publics can also look like panels to review procurement decisions made by government officials, public advisory boards to inform policy development, or oversight committees to serve as a check on authority and action of public bodies such as police. There are multiple modes of engagement in democratic spaces that afford ways to impactfully bring people together around data in support of participation, efficacy, empowerment, and justice.

Democracy and data

A key development in governance and data has been the growth of the *open data* movement, which pushed a conception of data as a key part of civic governance across the globe. If data is a resource, the open data effort focused on freeing it to empower innovation, market creation, and economic growth.[66]

46 CENTER IMPACT

In the US the open data push was driven by members of the computer-based free software and free culture movements, galvanized by a 2007 meeting in Sebastopol, California (just a two-hour drive from Silicon Valley).[67] A central tenet of their proposal was that public information should be published online in machine-readable formats and be freely available by default.

Their successes, and the work of many others, led to the passage of the 2009 Memorandum on Transparency and Open Government, which codified a US governmental policy of disclosure and data availability. Since 2012 the Open Data Institute has been central in rolling out this type of policy internationally, helping governments across the world create open data programs and a shared technology base to build on. Supporters point to stories of how open data has helped improve governance, created opportunities for innovation, empowered citizens, and helped solve real world social problems. Historian Theodore Porter argues that the generation of data for governance in many ways makes large groups of people governable; that modern society's approaches depend on this quantification of communities.[68]

Another more recent impactful development has been the rise of data-connected algorithmic processes in governance settings. A 2022 report from the Data Justice Lab at Cardiff University laid out negative results but also highlighted *alternative imaginaries*—new mechanisms for popular engagement that bring people together around data in support of democratic governance.[69] These alternates position civic data as a common good, produced by and belonging to all of society. Many software packages and services with civic intentions have been created to realize this vision, from systems that allow new forms of participation, to organizing tools that connect and amplify collective movements. Datafication has also transformed public procurement strategies, via combining massive open data records with information on corporate practices to leverage purchasing power for equity and justice goals. Another alternative is the idea of *Big Data abolition*—tearing down current structures and rebuilding new data systems that privilege resetting long-held power imbalances instead of reinforcing them.[70]

Those of us working with data in community settings are well served to ask how we can make every part of the process more democratic, creating impacts during both the process of working with data and impacts that come from the production and use of datasets. The core tenants of sovereignty, equality, and pluralism in representative democracy are guiding lights to keep in mind.

CSOs invite collective action

What image comes to mind when I share the phrase "open government data"? If you work in non-profit settings, that image might be a web-based data dashboard and detailed tables whose rows go on and on. To introduce what we can learn from CSOs I want to conjure a different image of open data in your mind—picture a group of school children counting textbooks in a rural Filipino village.

Textbooks are a critical piece of education across the globe, especially in areas that have low internet access. Major global attention and funding for textbooks came in the wake of an influential 1978 World Bank report.[71] More recently the United Nations Educational, Scientific and Cultural Organization (UNESCO) declared that "equitable access to high quality textbooks and learning materials is viewed as essential to successful learning throughout life."[72] With this context it comes as no surprise that in the early 2000s textbooks moved up the list of priorities for the Department of Education (DepEd) of the newly transitioned Philippines government. At the time, the best estimates indicated that only 60% of procured textbooks were delivered across the country; over a third of the textbooks the federal government had paid for never showed up to schools. This was leaving vast groups of children behind.

DepEd decided to partner with community organizations to help solve the problem, and the Textbook Count project was born in 2003. The partnership took an approach called *vertically integrated policy monitoring*—engaging with organizations at every point of the chain from government procurement to product delivery.[73] A key observation was that there was no feedback mechanism to demonstrate that books had or had not been delivered. Groups like the citizen-accountability focused G-Watch engaged the problem from a participatory perspective. They recruited thousands of citizens, teachers, and students to fill out standardized data sheets whenever textbooks arrived. These sheets recorded the location, date, and number of books delivered to schools. Aggregated reports by geographic region were then sent back to DepEd.

Beyond the production of the numbers, the rather photogenic process turned into local media spectacles, drawing news organizations looking for pictures of students counting the books. Driven by this data creation and public monitoring the procurement vs. delivery discrepancy started to fall quickly.

48 CENTER IMPACT

These children, teachers, and others were organized by groups to produce data about a government process that impacted their lives. They practiced what communications scholar Michael Schudson calls *monitorial citizenship*, creating data streams that hold institutions and elites accountable
for their shared responsibilities to society.[74] The case study has become
a go-to example of citizen-driven data production in the service of monitoring government procurement. Textbook Count offers a new example of
how to center impact for those working with data in any pro-social setting. The success was one driven by citizens taking control of datafication
and bending it toward their needs and goals; it was a redistribution of
power (defined here as capacity for self-determination in individuals and
communities).

The origins of CSOs

Of the multitude of terms used to describe groups that organize, act, and advocate, I find civil society organizations (CSO) to be the most helpful for this
book.[75] As defined by the United Nations (UN), CSOs are one way "people organize themselves to pursue shared interests in the public domain."[76]
This covers co-operatives, not-for-profits, non-governmental organizations,
labor unions, community-based organizations, and others. CSOs exist outside of corporate and state entities, fighting for rights, influencing policy, and
operating in support of community interests.

CSOs focus on a wide variety of social topics, including the protection
of freedom and rights, supporting socio-economic development, encouraging civic participation, offering accountability and transparency, and social
cohesion. These translate into actions such as organizing volunteers and members (like Filipino schoolchildren counting textbooks), producing events (like
a protest of Mexican mothers), convening cross-group dialogs, producing
reports and briefings, directly advocating to governmental powers, collecting
supporting funds, providing needed services, and more.

The core methods with which CSOs operate are broadly participatory,
inclusive, place-based, and impactful. While in practice some do become
power centers, they generally are focused on distributing and providing power
to the constituencies they are composed of. CSOs focus participatory power
onto collective problems like a glass lens might focus light, offering a spark
that grows into a larger fire for action and social change. CSOs bring people
together to create impact.

CSOs and data

CSOs exist to help their constituencies, those in need of services. Doing this work requires them to connect marginalized groups to the centers of collective decision-making, such as the government. With governance becoming more datafied, CSOs must speak that language of power to serve their populations effectively. However most don't have any formal training in the language of data. They frequently work with information in other forms but have found themselves needing to build expertise in the methods of quantitative data to engage with structures of power. This includes data production and data advocacy.

If CSOs bring people together for action on problems, and data is a way to understand a problem, then CSOs bring people together around data. Putting data out into the world doesn't immediately turn it into a community resource that is ready to use. Local CSOs play a critical role in engaging and educating their communities in what data is available to support their cause and how it can be leveraged for positive and negative impacts. The Detroit Digital Justice Coalition's *Opening Data* reports are a strong example, offering two handbooks with real-world instances of data discrimination, community data to support organizing, workshop facilitation guides, and creative approaches to data storytelling.[77]

The Textbook Count example serves here as a "textbook" case of this kind of collaboration to produce data in a pro-social setting (forgive the pun). If you work with data in community settings, think about how the data itself can be used to bring people together around shared goals.

Own the impact of your data stories

I'd heard the history of the town of Agloe, New York from a few friends, but didn't truly believe it until I looked it up myself. In the 1930s the founders of map-making giant General Drafting were concerned about plagiarism of their main map products. Following a trick of the time, they added a *copyright trap* to their map, drawing in a made-up town they named Agloe about two-and-a-half hours north of New York City. Anyone copying their map would be caught red-handed if they included this fictional town.

Decades later their plan came to fruition when map-makers Rand McNally included Agloe on a map in the 1950s. General Drafting executives pursued a legal case against their rivals for copyright infringement but found a surprising

50   CENTER IMPACT

result: their invented town of Agloe existed. There was an Agloe General Store at the precise spot on their map, and it was registered in the local Delaware County record as an official town.

Based on this curious history, Agloe has played a role in a variety of books and movies since then, offering a plot, point of intrigue, and mystery to novelists.[78] For those that work with data visualization, it reminds us that data has the power to structure how we see and interpret the world around us. We make data, and data makes us. Agloe came to life because someone built a general store, looked on the map of geographic data, and decided to name it after the town name they saw at that location; the map-makers couldn't have foreseen that impact of their invented data.

This power speaks to larger questions of data and mapping as a tool for being seen and gaining power (in non-fictional settings). Indigenous communities across the globe are facing constant threats from corporations and others that continue to encroach on their lands, even when officially recognized. They've responded in many places by repurposing digital technologies to survey and map their territories.[79] This intentional datafication of their ancestral land-rights, often with support and training from CSOs, is an example of bending the inherent power and technologies of data by communities toward their needs.

The history of data production and use aptly demonstrates legacies of negative impact. The histories and goals of data production and use in libraries, journalism, educational museums, representative democracy, and CSOs all offer goals and practices for creating an alternative. Learning from, and applying, methods and theories of the pro-social sector when working with data in community settings is a strong way to create positive impacts with data.

2

Focus on Participation

Existing theories critical of our dominant approaches to using data show us a path more aligned with pro-social sector goals. They reveal how participation must be a key focus for any community data project. Collaboratively designing and painting community data murals, and related work to show data in public space, expands the default toolbox for telling data stories to focus more on participation. Leveraging arts-based approaches to bring people together around data in community settings puts the justice focused critiques of datafication into action.

A data mural in Philadelphia

On a rainy Labor Day holiday in 2012 a few hundred Philadelphia residents gathered under a large tent not too far from the main historical center of the city. This crowd braved the weather to help the last stages of creating a giant collaborative data visualization—a three-story high mural on the side of an old warehouse. The community members represented all different ages, cultures, and neighborhoods, but they had all participated in the data production and analysis that led to the mural design. People came out on the gray day to push the year-long data project over the finish line. This group was working together on *How We Fish*, a project I describe as a qualitative community data mural.

Public murals are part of a rich tradition of beautifying space, amplifying political speech, sharing hidden or forgotten stories, and engaging people in participatory arts. One particularly relevant history is the growth of Mexican muralism, which emerged in the 1920s after the Mexican Revolution.[1] Artists of the time were advocating for art in public places that depicted stories relevant to the everyday lives of all Mexicans, making specific efforts to include marginalized groups such as rural and Indigenous peoples in their work and their visual stories. The results went on to inform the US-based Chicano Mural Movement in the 1960s, which accompanied grassroots mural creation with training workshops and explicit New Left political agendas and organizing.[2] The mural practices were a central tool, demonstrating the power of the arts

Community Data. Rahul Bhargava, Oxford University Press. © Rahul Bhargava (2024).
DOI: 10.1093/oso/9780198911630.003.0003

52 FOCUS ON PARTICIPATION

to organize, advocate, and disseminate messages in community settings. If you want to engage people from diverse backgrounds in some kind of participatory process, you're best served to look to the arts for inspiration.

Philadelphia has kept this long history of impactful community mural-making alive, driven by the Mural Arts Philadelphia project and its founder Jane Golden.[3] Their legacy is the country's largest public art program, with over 4,000 murals painted. Visiting the city, you can't help but notice that it is covered in murals everywhere you turn. Since 1984 the group has been using public murals to heal, celebrate, and build community across their city. They've partnered with the city, community groups, schools, businesses, incarcerated people, nonprofits, and more. The program's efforts focus on outcomes beyond beautification and urban renewal, encompassing positive impacts on education, the environment, health, and restorative justice across the city. An impact assessment study in 2003 found that the Mural Arts Project

> holds a unique opportunity as a bridging institution—to mobilize networks and to connect grassroots and community organizations with regional resources, government agencies, and private grant-makers.[4]

In the fall of 2023, I went to visit the mural that participants had worked on that rainy Labor Day holiday just over ten years earlier. I started near the grassy urban mall full of historical sites, turning west off the main sidewalk and toward Chinatown. Slightly lost but trusting my phone directions, I took the next right on N 8th St. to continue my search. Finally, I circled around a decaying abandoned police headquarters and the mural stood before me. *How We Fish* is hard to miss, covering the building like an elaborate quilt (Figure 2.1). Blue fabric melds with silhouettes of gears and trains, farmland sits next to a Chinese restaurant sign, and on top of it all is the text "Work Unites Us." The whole piece is a beautiful mix of glass mosaic and painted surfaces, lending it motion and vibrancy.

How We Fish was led by Social Impact Studios (SIS) and Eric Okdeh, beginning with the simple prompt to reconsider what the idea of work means to you. It was conceived of in the early 2010s, in the aftermath of a banking crash and the emergence of the Occupy Wall Street movement. This was a time ripe for reconsidering what roles work plays in our lives. In an interview, SIS director and mural co-creator Ennis Carter described to me how they wanted to understand the "opportunities that [work] gives us for independence, freedom, lack thereof, and agency." Another inspiration was SIS' *Posters for the People* research project, which collects the posters made during the US New Deal era

Figure 2.1 "*How We Fish*," a qualitative data mural in Philadelphia.
Credit: author.

Works Progress Administration. Carter describes that collection, and *How We Fish*, as "art as a means of celebrating workers."

To engage the city's residents in these questions, the team designed a participatory process built on facilitated conversations with various communities. SIS leveraged existing community partnerships, and built new ones, to recruit participants. At each workshop they asked residents to write what work meant for them on pieces of treated canvas, an accessible and popular material that Mural Arts often uses. They recorded quotes from individuals to produce data with the community itself. Carter pointed me to how each gathering offered local food and took place in local venues to ensure residents felt "valued and comfortable to share their story in a place of trust." Once the conversation sessions were completed, this qualitative dataset was analyzed to pull out key themes to represent in the mural, which were then translated into imagery and symbols by Carter and Okdeh. They focused on the symbols of work present in the language shared by residents; Carter was inspired using symbols she saw first-hand on a trip to see the Mexican muralists' pieces. Representative quotes were embedded back into the design, honoring the raw data the themes were pulled from—"work means teamwork," "work is more than making a living," "you have to work really hard if you care."

This mural showcases a data driven story about how work unites us. From my perspective the project wonderfully captures the idea of using qualitative data production, analysis, and synthesis to inform the design of a data visualization, all embedded in the empowering participatory practice of painting public murals. Carter approved of my labeling the project a data mural, describing their process as "making sense of ideas in some form that can be

54   FOCUS ON PARTICIPATION

transmitted." The end result? A huge data mural painted in public space to be viewed and considered by all who pass by.

The crowd in the tent that rainy 2012 Labor Day holiday was working on panels to be mounted on the wall. Many had participated in the data production workshops; some had their quotes directly in the mural. The local Department of Labor head gave a short speech and brought several staff to participate. While young and old painted, groups reminisced about the discussion they had about their history and relationship to working. Others watched slide shows from the past discussions, where people were holding up signs describing what work meant to them. They were socializing, but through these processes they were also actively engaged in critical reflective discussions about the data they had collectively produced, and how it was being presented back to them in the design of the mural. Video interviews from the event capture one local resident noting how "most of the senior citizens ... got a chance to remember how they came to this country;" a member of the facilitation team commented on how "people could identify: 'Wow, that came from me.'"[5]

It is easy to focus on the visualization itself, the mural is stunning. However, these quotes and the context of that activity show how the visual was just one output of the larger process. The entire data arts project used participation to reinforce community connections, spark thoughtful reflection and discussion, build new collaborations between organizations, and inspire and delight both participants and viewers. This was all built on the participatory nature of the data production, analysis, and visualization.

How We Fish builds on the tradition of using murals for community engagement, empowerment, and activism. The art form lends itself to settings associated with justice for marginalized groups in a variety of ways. Muraling doesn't require formal literacy, opening it to a broader set of people. It is also cheap: paint and labor do not cost much for a small team. The public context allows for large sets of people to see the messages in the murals. Murals can help enhance community identity and pride, improve mental health and well-being, drive new conceptions of the issues and topics they depict, and sustain and revitalize the community.[6] Political and protest murals offer stories and context for understanding ongoing events, build shared identity for groups under attack, provoke outrage and action, and serve as symbols within larger movements.[7] Murals build *efficacy*; residents living near new Philadelphia murals show an increased social cohesion and belief that they can change things through collective action.[8] Participants in community murals become active in place-making and place-keeping, form long-lasting connections to other individuals and organizations, and show sustainable bonds to others.[9]

These outcomes are exactly the kind desired by pro-social sector activities. *How We Fish* inspires us to reconsider how the goals of collaborative arts and community building can inspire new forms of working with data, shining a light on how critical it is to focus on participation.

Participation as a design goal

I broadly describe data projects in three ways—measuring to improve practice, telling stories to spread a message, and supporting experiences to bring people together. The first covers business practices such as using data to measure operational efficiencies and assess changes (e.g., key performance indicators, dashboards, etc.). The second is about organizations of many types using data when showing their impact and success to internal and external audiences (e.g., infographics). The third is where most of my interests lie, and the one I think is most under-explored: creative approaches to using data to bring people together for community building, deliberation, and action.

Designers and analysts model data storytelling in a variety of ways[10]—I've seen circles, braids, inverted pyramids (from journalist Paul Bradshaw[11]), and overlapping hexagons (from the School of Data). Whatever shape you use to depict this process, each step of working with data offers opportunities for participation. I'll use my own simplified workflow for data storytelling (Figure 2.2) and the *How We Fish* project as an example to tease out how this can look in practice.

While asking questions you can encourage participation to ensure that a diverse set of perspectives on the data are included. The questions about data surfaced by a group with one set of norms will likely differ from those highlighted by another. One approach to fight against this tendency is to reach out to those groups most represented or impacted by the data to solicit ideas and narratives they are curious about. The *How We Fish* team started with a question and theme from their own perspective—work—but quickly moved to asking others to share their thoughts.

Figure 2.2 My simplified workflow for moving through a data storytelling project.
Credit: author.

56 FOCUS ON PARTICIPATION

The step of producing data easily lends itself to shared practices. Human-centered data production processes are best achieved by those who are trusted by the subjects of the data; this may or may not be you. Engaging with various groups to help collect and validate data is an easy way to make data production more participatory (think back to the Textbook Count project I introduced in "CSOs invite collective action" in Chapter 1). Citizen science projects show us one approach to how creating data can focus on participation and value "just good enough data."[12] The *How We Fish* team hosted multiple workshops to produce a qualitative dataset they could use to guide the design of the mural narrative pieces and visuals.

Each dataset contains a myriad of stories that can be told. You can build practices that introduce the data to various communities to see what potential stories emerge from their unique perspectives. Civic data analysis can involve diverse community members. The *How We Fish* team surfaced themes from the qualitative dataset themselves, but SIS was itself a member of the community and thus had embedded trust, adding to the long-standing relationships and reputation of Mural Arts Philadelphia.

Finding an appropriate way to tell your story must center on the context and goals of a project. Who better to judge this than the very audiences you have in mind? Their participation can point toward the potential success of the method you choose for telling a story. Picking the format for representing a data narrative can be a collective process. The *How We Fish* team started with the idea of a public mural but took visual design cues from the narrative themes that emerged from the community conversation sessions. More practically, many sections of the mural itself were created by those community painters and mosaicists of all experience and ages.

Trying out the data story you have created is the final step described in my process. This, again, lends itself to participation with the audiences you have in mind. The community response to *How We Fish* designs at the final painting and mosaics session is an example of getting this type of feedback.

Inspirations from pro-social settings

It is easy to say we need to rethink our data processes around participation; it is harder to envision what that looks like. The word data generally suggests images of statisticians, spreadsheet-based analysis, and chart-making software. However, there are broader ways to think about data that specifically relate

INSPIRATIONS FROM PRO-SOCIAL SETTINGS 57

to its context of use and social impacts. Examples from data journalism and urban planning flesh out new ways to think about participation in practice.

Participatory data graphs

The *New York Times' You Draw It* series offers us an inspirational example of how to use participation in creative ways online. Each of the three data journalism articles in the series begins with a partially drawn line chart; the reader is invited to draw how they think the curve continues. This reader-authored speculation about the dataset is then used to lay out the core message of the article and associated dataset.

The first 2015 article asks you to draw a line representing the correlation between the percent of children attending college and parents' income percentile (the central datapoint is offered as an anchor).[13] Once you've drawn what you think the curve looks like, the piece scrolls down and shows you the true curve and some dynamic text that summarizes what you got right and wrong. Continuing further, it shows a heatmap of reader responses overall. The net effect is to allow comparing individual, collective, and evidentiary modes of understanding the issue. Most readers underestimate the number of lower income children attending college, and over-estimate the number of high-income children. This creates an impactful nuance to the story, building on the participatory invitation to add to the dataset and create a comparison with the reader's perceptions. The impact is clear: it forces me to draw my perception and then offers a subtle form of correction, leaving me with a new understanding based on rigorous data about the issue. Beyond that, it pushes me to consider the potential impacts of the collective misperceptions about the issue.

The second piece, from early 2017, focuses on comparing the differences in metrics on key socio-economic issues during the Bush and Obama administrations.[14] It includes a series of charts showing the trend over time for a selection of issues during the Bush years (unemployment rate, deported immigrants, spending on health care, etc.) and asks you to draw how you think each progressed under the Obama years. After each is a "Show me how I did" button that reveals the actual data and a short description of it. Unlike the first piece, this is no other narrative included besides the seven interactive charts; the whole piece is just asking you to "see if you're as smart as you think you are" (the subtitle). The entire narrative is built to assess your perception and compare it to the official metrics.

58 FOCUS ON PARTICIPATION

The third in the series, from later in 2017, is narratively a mix of the two prior pieces.[15] It opens by asking readers to complete the line chart on four graphs, American deaths due to car crashes, guns, human immunodeficiency virus (HIV), and drug overdoses over the last fifty years. The final chart rises exponentially since 1990, clearly standing out from the other three and setting up the topic of the text that follows.

Taken as a series, the core participatory aspect is the invitation to complete a chart (or charts) at the start of an article. Sounds a bit mundane, but I encourage you to visit the articles and do it; in action the impact is to make you pause and think. You are forced to consider your own preconceptions of the issue through the act of estimating the data visually; priming you to be receptive to further graphical representations of it, and potentially even opening you to being corrected. Arguably that last point is the narrative design hypothesis of the authors that led them to use the interactive chart drawing activity. The designers have effectively engaged a participatory interaction (drawing) to a traditional form (the graph) to pull the reader into the story and connect them to the data. Restricted to well-known forms their readers are used to and expect from the New York Times, the authors nonetheless pushed at some of the rules to create a novel approach that can inspire us to rethink what we are and aren't allowed to do within the norms of standard charts.

Participatory urban planning

Urban planning has long-established practices for bringing people into community civic processes. Two examples from different parts of the globe demonstrate this concretely.

The first is from my own city—the Go Boston 2030 planning process. This participatory planning project was designed in 2015 by the Interaction Institute for Social Change (IISC) for the City of Boston to develop a vision for the transportation future based on outreach and community dialog.[16] IISC is an organization that focuses on building the capacity of individuals, communities, and organizations. Created based on a theory of going "all in" on democratic engagement, the campaign rethought data-centered public participation.[17]

The key component was the Question Campaign, a process used to invite Boston residents to ask questions about how they would move around Boston in the future. Advertisements pushed the prompt across the city and the team solicited questions via text, email, handwritten cards, Twitter, Instagram, and more. For in-person events they even built a customized glass-sided truck to visit each neighborhood and source questions on paper. Throughout 2015 they partnered with the city and local groups to make sure a diversity of voices was engaged.

This created a qualitative dataset of questions sourced from the very population their planning decisions would impact most—walkers, bus riders, bikers, and drivers. In most data-centered planning processes this data would then get turned over to an analyst for labeling, coding, and evaluation to produce a report back to city leaders. IISC put a different approach on the table. To process the 5,000 questions, they led a half-day event that invited community members to review the questions and generate a summary of key themes, very literally inviting them to have a seat at the table (Figure 2.3). This again centered the participation of the community, specifically in refining the solicited knowledge embedded in their data (rather than handing it over to a third-party domain expert). Participatory data analysis often looks like a room full of people chatting around circular tables. Going further, IISC then created an event to reflect the findings that the sessions had gleaned and articulated back to the community for verification and validation. They also published a report with the findings, but this event offered a parallel

Figure 2.3 Participatory analysis of the GoBoston 2030 questions dataset at an "idea roundtable" event.

Credit: Go Boston 2030, Boston Transportation Department.

60 FOCUS ON PARTICIPATION

checkpoint for community ownership of the story being told about this qualitative data. Each step of the Go Boston 2030 planning process was considered as a point to build participation, efficacy, and distributed ownership of the results.

A few years later, and halfway around the world, the Tanzanian group Data Zetu took a very similar participatory approach to their data projects. Working with development data in their sub-Saharan East African setting, they noted how communities are often approached as sources of data by non-governmental organisations (NGOs) and researchers, yet seldom is the data brought back to them. They concluded that this cycle "alienates the community from their own data and disempowers them from taking action based on their collective evidence," echoing harms of extractive colonial practices we still feel today.[18] A growing set of practitioners and academics label this phenomenon as *data colonialism*—"normalizing the exploitation of human beings through data, just as historic colonialism appropriated territory and resources and ruled subjects for profit."[19]

In 2018 Data Zetu's response was to create "share-back sessions," where they presented data back to the community and facilitated a discussion with community leaders about the challenges it describes. This put the data very directly in service of community goals, resetting the power imbalance so often standard in extractive data projects. Conventional projects collect data from a community only to analyze and discuss it in some other setting far away from the subjects and distanced from any impacts that might be felt. Operationally, the Data Zetu team synthesized a "community insights" document summarizing the cleaned and analyzed data. These documents were used as inputs to civic meetings, and introduced the very idea that data could be produced to understand a critical community challenge.

The Data Zetu team had a few key findings in their first year of this using this practice, each demonstrating the power and potential of their participatory share-back approach.

- Data created by community members helped catalyze dialog and debate.
- Sessions generated requests to help manage data that communities were already producing.
- Conversations surfaced root causes of the challenge not exposed by the data itself.
- Participants showed an increase in valuing data to inform goals and decisions.

Data Zetu describes this part of their work as "borrowing the data," and the share-back sessions offered them a way to give it back. Participation was key to helping them rethink standard approaches to using community data for planning and action, moving toward decolonized data production practices more appropriate for their context. We use a similar process in our data murals (see "How to make a data mural" in this chapter), as do a growing number of other groups.[20]

These examples, from Boston to Tanzania, demonstrate how urban planning processes model creating points of participation in various stages of data projects. Go Boston 2030 shows us that participatory data experiences are possible to create, and critical to deploy, at all stages of data work and representation in the pro-social sector. The findings from Data Zetu show that a person's perception of efficacy increases when they are engaged in the doing of something, rather than simply being told it is possible; participants believe more strongly that they can make a change after a share-back session. To fully engage with this approach to participation we must push beyond the standard processes for engaging participation and remake some of the ways we think about data and the role it plays in our rapidly datafying society.

Critical theories of data use

Participation offers a way to align your community data project's goals with those of pro-social sector organizations. These on-screen and off-screen examples show some ways that can look, but we also need to flesh out the theoretical motivations. Scholars and activists have taken up this question and formulated more critical approaches to how we think about data in society. These alternative framings provide important guidance for resetting who gets to work with data, and how, in community contexts.

Data for good

A growing critique of the problematic history of data, and modern focus on data science in support of capitalist aims such as efficiency and profit, has led to the idea of intentionally using data for social good. The *Data for Good* movement specifically calls out modern corporate practices of data as not in service of the common good; data for good exists in opposition to these historical norms. Researchers, academics, and activists in this space have been laying

62 FOCUS ON PARTICIPATION

the groundwork for a socially responsible data science practice, with its own set of ethics, norms, and case studies. New York University data policy expert Anne L. Washington's 2023 book *Ethical Data Science: Prediction in the Public Interest* is the most thorough dive I've seen into the application of ethics in the field of public interest technology development.[21] Notably the work draws from methods in the arts and humanities, perhaps influenced by Washington's Master's degree in Library and Information Science.

Numerous types of groups operate in the data for good space. Data.org is a major example, created in 2020 by a joint US$50 million grant from the MasterCard Center for Inclusive Growth and the Rockefeller Foundation.[22] They primarily convene the field, disseminate thought leadership, and fund award challenges to push the sector forward. Nonprofit DataKind is another impactful example, working to bring volunteers together with nonprofits to tackle problems where data science skills can push forward mission-driven work. DataKind enlists volunteer data science experts from commercial businesses, and in parallel helps nonprofits define their data problems in a way the data scientists can understand. The resulting collaborative projects create data-driven solutions that align with the missions of the organizations. Nonprofit Data4Change operates with a very similar model (and was notably co-founded by a journalist). In the global development sector, Data2X responds to data-driven development processes by arguing for disaggregation of gender data to identify specific subpopulations that might be left behind despite overall improvements in metrics. These organizations, and others, inspire, ignite, and unite the now-established data for good sector around questioning who data is about, who owns it, and who stands to benefit from (primarily computational) data projects.

The data for good movement is spreading, driven by those development sector groups and academic practitioners seeking to focus on the social impacts of data science practices. "Data Science for Social Good" (or DSSG) is becoming a commonly known term in universities. Members of influential academic groups such as the Association for Computing Machinery (ACM) have created related tracks at annual conferences, and completely new annual gatherings such as the Conference on Fairness, Accountability, and Transparency (FAccT). These spaces build community and academic validity for the idea that working on data science in service of the social good is an acknowledged and accredited pathway within teaching, learning, and research in higher education.

The data for good label has come to be used so often recently that it is hard to come to agreement on a central definition and core goals.[23] Some have responded to the rapid growth with critiques that can inform iteration and further refinement of programs, including:[24]

- Volunteer-driven efforts are hard to sustain over time.
- The most attractive problems to volunteers are often the "cutting-edge" ones, rather than the mundane-but-impactful ones.
- Companies donating technologies or staff can market their service in a way that absolves them of having to question their central work and its harms to the sector at large.
- Partnering doesn't help build capacity in the underserved areas that receive the benefits of the projects.
- Separating data used for "good" suggests it is acceptable to use data not for good.

Acknowledging these potentially negative aspects is important but doesn't preclude holding up the positive impacts of the larger movement to put data science in the service of social good. Taken as a whole, the movement is building capacity for the typical subjects of data to control how it is owned and analyzed, even if they don't do it themselves. Data for good also provides pathways for data science learners to fully embrace responsible and ethical uses of data that address power imbalances, rather than simply approaches molded by profit-driven corporations.

Many of the critiques can be synthesized and described as part of a burgeoning effort to *decolonize* data practices (like the Data Zetu share-back sessions). Despite good intentions, working with data in development or non-profit settings often amplifies existing injustices and inequalities that are the leftovers of our global colonial history.[25] The European "Age of Exploration" led to violent land seizures, subjugation, and enslavement of vast populations, theft of riches, and other atrocities. Monetizing data extracted from subjugated groups is an act of data colonialism.[26] Data decolonization operates with a restorative approach, arguing that this legacy must be actively undone via conscious and intentional action on the part of the colonial powers.[27]

The movement to decolonize data, particularly in global development settings, centers a set of principles that require asking questions to understand for whom a dataset is being created, how to use the processes of producing and representing data to empower the community it represents, recognizing the context of the data, and questioning for whom and how data is being represented. More often than not, participation is a key aspect of processes in decolonial data projects. As one example, the Connected by Data

64 FOCUS ON PARTICIPATION

project argues for participatory approaches to policy development and data management, sharing case studies of participatory data governance in the real world to establish best practices and learnings for future projects.[28]

Indigenous groups are significant stewards of the movement to decolonize data, presenting an alternative to data for good that specifically seeks to dismantle the structural power embedded in modern data practices. The principles of Indigenous data sovereignty are encoded in a number of ways, perhaps most impactfully as the *CARE* principles for Indigenous data governance.[29] These specify to create data in support of *C*ollective benefit, give local communities *A*uthority to control that data, take *R*esponsibility for relationships and capacity, and follow *E*thics to minimize harm and preserve for future use. This set of guidelines is inspiring new approaches to working with data that push on assumptions central to data for good projects.

Data for good projects seldom engage with what has created the structural threats, or individual risks, that they respond to and seek to address. In many ways Data for Good projects focus on downstream issues, the impacts of some deeper problems that are happening upstream. In addition, the movement has close ties to the international development sector and can be critiqued in many of the same ways, specifically regarding power asymmetries and entrenched colonial thinking. Indigenous data processes offer alternate approaches grounded in groups actively experiencing harms from colonialism.

Critical data literacy

The theory of *critical data literacy* offers a helpful lens for digging deeper into these concerns, centering social transformation as the goal of data literacy (see "The concept of data literacy" in Chapter 3). Teachers are creating liberatory learning experiences related to data built on individual and collective development of critical consciousness. These teachers engage with the data trails we leave behind by asking students to question and examine the mechanisms and intent of the data.[30]

The Local Lotto project serves as a helpful example of how this can play out in practice.[31] Created by the City University of New York in Brooklyn, the Center for Urban Pedagogy, and the Massachusetts Institute of Technology (MIT) Civic Data Design Lab, the project introduced a data literacy curriculum into an under-resourced school in New York. The team designed their process to use data on a culturally relevant and local issue as an onramp to working with data: the public lottery. Run by the state since the 1960s, the lottery offers almost a dozen chance-based games for purchase, with all funds being put toward the public education system.

The first phase of Local Lotto led students through interviewing neighborhood residents about their participation in, and perceptions of, the lottery system (Figure 2.4).[32] Just as libraries often do in their data literacy programs, the team engaged learners as producers of qualitative data about this locally relevant topic, rather than consumers of some data from outside their community.[33] Learners also used tablet computers to produce quantitative data about lottery purchasing habits and sales volumes. The datasets described spending frequency, geographic locations, and motivation; in the process they built both interviewing and data creation skills.

The second phase of the project collated that data into an interactive set of digital maps to support their inquiry into the lottery's impact. Students were invited to contribute their own knowledge about their neighborhood's socioeconomic status. This led to immediate observations from students about

Figure 2.4 Students interviewing lottery players in their neighborhood.

Credit: Image provided by Sarah Williams, Civic Data Design Lab (MIT) and Laurie Rubel, Brooklyn College.

66 FOCUS ON PARTICIPATION

lottery spending being inversely correlated with income; low-income areas spent a higher proportion of their earnings on lottery than high-income areas. They also noted that on average purchasers spent far more on lottery games than they won; the data showed them that the lottery was more akin to a casino than anything else.

The third phase of Local Lotto pushed into application of this data in argumentative form. Introduced to provocative statements such as "the lottery is a tax on the mathematical illiterate," students were facilitated in a debate format to argue in support of or against the lottery. Many students pushed back on this prompt, arguing that the questions' framing set up a deficit-laden narrative they found uncomfortable and discriminatory. Driven by these discussions, students created multimedia narratives that used the data they had produced, combined with official state data, to support their points of view. Students built critical data analysis skills while constructing these data-informed arguments. They were debating the pros and cons of a system they had all seen or interacted with, in a local setting they knew intimately.

Local Lotto demonstrates many aspects of one of my inspirations, educator Paulo Freire's approach to building critical literacy skills in reading education. Working with marginalized communities in northeast Brazil in the 1960s, Freire argued that decontextualized literacy instruction was not sufficient for the needs of the illiterate and disempowered. In the Freirean approach, only by learning the context of what one reads can the learner truly understand what they read. This means reading about their social condition, social structures, and barriers to growth in sources like the news. Freire focused on developing *conscientização*/critical consciousness in his literacy education work, not just the raw ability to read.

Inspired by Freire's work, the Local Lotto team took an approach of critical data literacy to the activities they led students through. Participants knew what the lottery was; through the data they constructed an understanding of its multiple roles in society and wrestled with its positive and negative impacts. They interrogated who gains from it via data analysis, and the motivations of those that choose to participate by buying lottery tickets via data production. The youth constructed an extended dataset through the interviews and observations, building a context within which the existing official lottery data could be more fully understood from their lived experiences. They built a critical understanding of the data, linking the otherwise abstracted observations to their emergent and existing understandings of the social context of the lottery itself in their community. Local Lotto introduced data literacy just as Freire introduced language literacy, in service of empowerment and critical thinking within a social context.

Across these phases, Local Lotto was able to demonstrate students taking a critical perspective to producing, interpreting, and arguing with qualitative and quantitative data. Besides collective measures of confidence and comfort with data, they also showed more inter-personal outcomes such as one student using his work to convince his mother to stop buying lottery tickets.

In a systematic review of educational work related to critical data literacy prepared by the National Academy of Sciences, researchers found shared central tenets across projects that included:[34]

- Uncovering social inequality with data.
- Confronting the ethics of datafication.
- Building civic imagination with data.
- Individual expression with personal data.

Critical data literacy introduces the ideas that data production, analysis, and representation are best situated within the social contexts of the learner. Individuals build critical data literacy when they embrace that context and engage it in their understanding of the data they are working with.

These goals of critical data literacy flesh out key principles of addressing some of the upstream problems left unaddressed by approaches like data for good. They also reinforce the goals of pro-social spaces via participatory invitations to learn. Critical data literacy helps us think about why participation is so important to working with data in the pro-social sector. However, there is more to dig into when you zoom out to data's impact on the social structures of society. Critical data literacy opens the door so we can see structural harms and remedies within a datafied society, but we need complementary theories to flesh out how to conceptualize and address them.

Data justice

I had two hours left before a deadline and my power level was running low. I don't mean metaphorically; the battery of my cordless drill was almost empty. I frantically scrambled around the tables of the MIT makerspace I was working in, surrounded by mechanical and digital fabrication tools. Shiny metal surfaces glinted at me from all angles, and the sounds of hammering and cutting

68 FOCUS ON PARTICIPATION

bounced around the active two-story open studio space. I had another battery somewhere in here, and I needed it to finish drilling together the wooden structure I was building for the event starting in just a few hours.

In that makerspace I was building a physical data collection and reflection tool for the Data for Black Lives conference (D4BL), a gathering of organizers, technologists, students, and activists created to reclaim data as a tool for social change. Created by Yeshimabeit Milner and Lucas Mason-Brown, D4BL focuses on those most impacted by the negative harms of datafication in public settings. In the US, those burdens are borne mostly by the communities of Black people, Indigenous people, and people of color. D4BL hosts multi-sector gatherings, runs programs to increase positive impact on their member communities, fights to redirect policy toward restorative data practices, and works in media channels to shift the narrative around data science work that has social impact. They link current-day big data practices to a long legacy of capitalist technology used to suppress and subjugate populations.[35]

Through academic research and investigative journalism, we have ample case studies of ongoing structural harms that don't need revisiting in depth here.[36] The case is well-established at this point that datafication has reinforced, amplified, and created systemic harms and risks, and that those are not equally distributed across all groups of people in society. The concept of *data justice* helps us understand changes in data-centered social systems and their impacts for large groups of people. With this understanding we can reframe collectivist responses that move beyond individual issues such as data literacy and privacy to more collective participatory action.

Back in the cavernous MIT makerspace, I was still struggling to find batteries to assemble a pile of wood into the *Just Data Cube* I had designed, an open structure measuring eight feet on each edge. After a few minutes of scrambling, I found the battery and got my wooden walls attached and mounted in preparation for the day's events.

Three months earlier I had learned the inaugural 2017 D4BL conference was being hosted in the building where I worked at the time. Excited to get involved based on shared interests I perceived between our data projects; I sat down with co-founder Lucas Mason-Brown to brainstorm. The resulting idea was to inform their closing manifesto by collecting qualitative data from participants on their demands of themselves and others, and their commitments to the same. Surveys didn't really activate the space at public gatherings, so instead I proposed to construct a two-sided structure where attendees could write these demands and commitments on the walls.

Figure 2.5 D4BL conference attendees writing on The Just Data Cube.
Credit: author.

Over the course of the two-day event the participants imagined more just uses of data, an ecosystem focused on equity and justice, and captured their ideas by covering the walls I had installed (Figure 2.5). It collected a few hundred notes, which post-conference were collated and summarized as part of a manifesto from D4BL and informed their strategic planning. This participatory data production mechanism engaged people who might not otherwise fill in a survey. The act of writing in colored chalk on the black walls created an externalized physical memory of the commitment or demand. The object invited reflection and discussion as you read over what others had written. At its core, the Just Data Cube was an invitation to reflect on data justice collectively, sparking introspection and conversation using the techniques of participatory museum exhibit design.

Current work on data justice focuses on institutions and organizations that play outsized roles in the systems of datafication, questioning the power dynamics and paradigm they bring to play. D4BL is just one organization that

70 FOCUS ON PARTICIPATION

works on these issues. The Detroit Community Technology Project includes data justice as one of the three pillars of their education and advocacy work, offering participatory activities and facilitation guides to engage the community in critical questioning about impacts.[37] They also facilitated two data murals at local schools in order to "spark conversation within the broader community."[38] The Data & Society research organization includes the data justice lens as part of a larger agenda that focuses on issues from artificial intelligence (AI) and automation to datafication in health, labor, and more. The more directly named Data Justice Lab (hosted at the Cardiff University School of Journalism, Media and Culture) runs research projects focused on the impacts of these data structures and systems around us. They also host an annual convening to bring the academic and civic community together.

At its core, the concept of data justice lives at the intersection of datafication and social justice. In my work I define *social justice* as a social structure where members of society can participate fully in their roles, achieve social mobility, and access collective support structures provided by their communities. Feminist philosopher Nancy Fraser describes the participatory equality this allows for as the "social arrangements that permit all to participate as peers in social life."[39] Data impacts social justice through a variety of means, used both intentionally and unintentionally to create harm for vast sectors of society and limit various groups' capacity for achieving equity.

The Cardiff Data Justice Lab describes the work to be done across a variety of approaches.[40] Civil society groups and technologists in the sector often overlap in their fights for justice within state systems, creating an opportunity to bridge between the two. In the development sector there are widely shared and invested-in principles that can be leveraged to understand transnational approaches. The influential 2014 United Nations (UN) *A World That Counts* report argued that development agencies shouldn't be allowed to say "we don't know" in a datafied world.[41] While I criticize this statement for echoing the myth of quantitative data as all-knowing facts, and ignoring experiences and information that can't be quantified, the idea speaks to how widespread datafication has become. Existing rights frameworks—migrant rights, labor rights, anti-discrimination, etc.—can be brought to bear in datafied systems to offer a language of opportunity and infringement. Alternatives can be imagined to design systems that use data hand-in-hand with the groups most impacted. Computer scientists are increasingly exercising their "right to refuse" to design systems that further centralize power or create active harms in marginalized communities, such as technologies for separating migrant families.[42]

CRITICAL THEORIES OF DATA USE 71

Work from these organizations, and others, pushes us to reconsider what social justice looks like in the age of widespread datafication, focusing on the structural inequalities perpetuated and created. What role should and shouldn't data play in our society? What impacts does it have on equity, fairness, and power? These are collective structural concerns.

One category of systems data justice questions is those where datafication has created non-democratic processes in otherwise democratic settings. These are systems that should be democratic but are sold as private opaque solutions—predictive policing, housing access, child welfare support allocation. In the US these projects often embed data that reflects historical structural racism and segregation, and unsurprisingly produce outcomes in that vein. Another set of examples can be found in the targeted advertising of political messages. The 2018 Cambridge Analytica/Facebook disclosures revealed how normal operating practices of digital advertising giants created opportunities for micro-targeting of messaging to voters modeled as receptive based on their online behaviors. Democratic systems of collective organizing and governance are being co-opted by non-democratic datafied processes.

Data justice highlights how data is being used in social settings that impact discrimination, power imbalances, and social justice. What do we demand of our structures? What do we commit to create as alternatives? Individual-level responses and participation are necessary preconditions to responding to these harms but are perhaps not enough. A more collective participatory and systemic policy response is needed to complement that.

Data feminism

Data justice work in general leaves unaddressed how our methods for working with data, and representing it, connect to these collective systemic harms. The approach of data feminism gives us more concrete approaches to take in community settings, built on a foundation of decades of feminist thinking about equity, inclusion, and related issues that are critically important in those contexts.

Feminism has a long tradition of questioning power structures that suppress social, economic, and political equality across society. For our purposes I'll define *feminism* as the practice of labeling and fighting against oppression, such as sexism, with the goal of creating a more just world. Its history is often described by a series of "waves" in the West.[43] The first wave was driven by civil society movements to secure women's right to vote in the nineteenth and twentieth centuries. The second wave focused on liberation, seeking legal and social equality for women from the 1960s and beyond. More contested definitions follow, with a third wave in the 1990s perhaps focused on questioning and reimagining depictions of women across media, and a fourth wave of the 2010s perhaps focused on empowerment via activism and intersectionality. At various points the movements have wrestled deeply with questions of shared definitions, inclusivity, external mockery, and more. These waves, however one defines them, have produced significant levels of activism, theorizing, and writing that continue to influence generations of people working for social good and equality. They offer novel approaches to participation and working with data from outside of mainstream practices.

From a feminist perspective, visualization norms reproduce biases and inequalities in society.[44] Data visualizations embed political and ethical perspectives. The perceived objectivity and rhetorical power of science and charts lend visualizations a veil of authority and correctness and can reproduce power asymmetries. *Data feminism* names and describes those disparities, connecting them to our methods of statistical analysis and visual representation. It pushes us to take further steps toward justice, even if the overall task feels daunting.

D'Ignazio and Klein deeply explore the implications of feminist thought for modern data practices in their impactful 2020 book *Data Feminism*.[45] They offer the reader a set of seven guiding principles.

1. Examine how power operates in the world.
2. Challenge unequal power structures to work toward justice.
3. Elevate human emotion and embodiment over the myth of rational objectivity.
4. Rethink traditional binaries and hierarchies encoded in classification systems.
5. Embrace multiple perspectives on knowing something.
6. Consider the context of data to truly understand it.
7. Make the labor of data projects visible.

These principles echo many of the approaches I take, relating deeply to ideas and examples throughout this book. This should come as no surprise—I first met and worked with *Data Feminism* co-author Catherine D'Ignazio while we were both at the MIT Center for Civic Media. Our collaborations around data pedagogy, literacy, and public interest technology began there, and have continued in the decade since. I've been deeply influenced by the approach of data feminism. Specifically, it offers us a lens for understanding more of the upstream causes of systemic data harms.

With this feminist perspective in hand, numerous projects have taken up the idea and practice of data feminism and used it to describe and strengthen their work with data.[46] This is fertile ground with much to explore, and this book and my own work could be accurately defined as approaches to practicing data feminism. It is one of the set of theories that inform new methods to add to the toolbox for appropriate and effective data storytelling in community settings.

Paint data murals

Data for good, critical data literacy, data justice, and data feminism offer us richer ways to think about why participation matters in community data projects, and the role data plays in society. They are some of the bedrock theories that connect to my conception of popular data. One key to expanding the toolbox for practitioners looking to apply these ways of thinking in the field is to bridge from ways of thinking to ways of acting. What are the new techniques that we can use with people in community data contexts that implement these theories? For this, we can return to *How We Fish* as an evocative example of the idea of public displays of information presented in low-tech ways. Data murals offer a participatory process that is a new tool on your belt for bringing people together in service of the goals of critical theories of datafication.

Inspired by the history and power of public community murals, since 2011 my wife Emily and I have been iterating on a systemic and evidence-based approach to creating empowering and impactful *data murals*. These are facilitated collaborative processes that move from data to a story, to a visual design, to a collaboratively painted mural in public space. Data murals flip the script on historical extractive data processes, offering participatory methods for bringing groups together to interpret and tell data stories about themselves. This puts the power of analysis and storytelling into the hands of the traditional subjects of data.

Painting data on walls

The years adjacent to our first data mural saw a burst of activity connected to data and displays on public surfaces. Artists, designers, and activists were exploring the idea of painting data in public space to activate engagement, insight, and empowerment.

In the same year as our first data mural, 2011, Jose Duarte began using simple craft materials to create temporary bar charts on walls in public spaces around Bogotá, Colombia. He created a series of pieces about how Latin American citizens thought of their cities; picture a bar chart made out of paper taped on a brick wall in a park.[47] Duarte's goal was to bring this data back to the streets that it represented, allowing citizens to interpret the aggregate results themselves and connect them to their lived experiences.

In June 2011, influential new media artist Golan Levin was playing with the idea of data graffiti. In a quick project sprint, he created an adjustable laser-cut stencil that made it easy to quickly spray paint pie charts with descriptive text (Figure 2.6). The clarity of the black-and-white chart, and the visual aesthetic of tagging, lent this work an insurgent and counterculture feeling, reimagining who is allowed to work with data and in what settings.

Across the ocean in Brighton, UK the year prior saw residents there collecting data about their energy use as part of the 2010 Tidy Street Project. The university team helped residents track energy usage and then worked

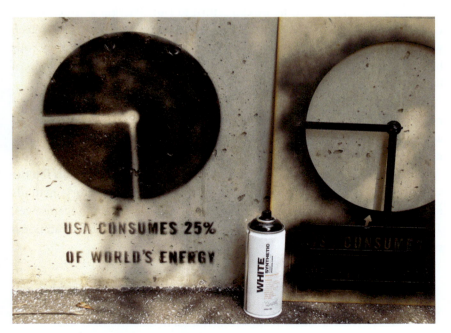

Figure 2.6 Levin's "Infoviz Graffiti."
Credit: Golan Levin.

Figure 2.7 Public visualization of average energy usage for residents on Tidy Street.

Credit: Jon Bird and Yvonne Rogers.

with a local graffiti artist to visualize the aggregate data in public form, on the streets of the city for all to see (Figure 2.7). They found this collective visualization created a sense of community pride in having reduced their collective energy consumption.[48] Participants took specific actions to change their patterns of energy use and were pleased when they saw personal and collective reductions. The integration of the artistic display connected directly to feelings of ownership of the change and affiliation with the community.

Also in 2010, artist Sebastian Errazuriz was struck by the caskets of US soldiers being unloaded at airports (victims of the ongoing invasion of Iraq). Driven by concerns for the mental health of those who survived, he soon found that far more committed suicide in 2009 than died in combat. He decided to bring these unseen numbers to larger audiences, grabbing black spray paint and creating a simple tally based mural outside his New York City studio (Figure 2.8). This public display of a simple comparison of two numbers was easily understood, surprising, and shocking. Through its public form his piece forced people to see and acknowledge the problem.

Taken collectively, these pieces offer a glimpse into the wave of public display of information on walls and streets around the world around 2010

Figure 2.8 Errazurriz's tally-based mural comparing deaths among US Soldiers in 2009.
Credit: Sebastian Errazuriz.

and 2011. Notably lacking are examples that are widely participatory in their creation. A handful focus on questions of empowerment, but others are more typical data processes with an alternate final representation. What does a participatory approach to creating data murals look like? How can the key principles of popular data inform and establish data murals as a normal practice in social settings that are data centered?

How to make a data mural

Since 2011 my wife Emily and I have iterated on an approach to creating data murals that is built around four project phases: story finding, visual design, painting, and presentation. For us the data murals are as much process as product.[49] The act of co-creation with the community engages experiential ways of connecting to the data, pushing past abstract representations like charts and graphs. We've painted about a dozen data murals around the world, and our work has directly inspired many others to create their own.

To flesh out the process of creating a data mural I'll return to the one we created with Lisa and the Collaborate for Health Weight Coalition (C4HW) coalition (as described in the "Introduction"). Supported by a federal program, this multi-sector collaborative had collected data focused on three to five-year-old children at local Head Start family service centers.[50] Using community-based participatory research(CBPR) approaches, the project staff had been working to improve collaboration between staff, parents, community leaders, and the government. Some of their outcomes included better access to physical activity and increased data tracking related to health. As part of their dissemination and engagement efforts, they invited us in to lead a data mural process with these same groups.

The initial design session brought together stakeholders (parents, staff, and civic leaders) around a multi-page data handout. Working through a variety of our hands-on activities, participants moved from standard tables and graphs we shared with them to a one to two sentence story they wanted to tell in mural form.[51] To support the story development, we offered a narrative template: "The data say_____. We want to tell this story because_____." For the C4HW project, that full story read:

> [t]he data say that the economic status of families, cultural traditions, and where you live play a part in the health of all children. We want to tell this story because the culture of the home, the culture of the community, and the culture of its organization needs to work together towards improving child health.

The data analysis and story generation were all participatory and engaged a set of people with varying levels of self-assessed data literacy. Narrowing down on a single story to tell was challenging, because participatory processes that honor multiple perspectives are so often messy. The dialog around the data opened doors into deeper conversations about how the services worked, what role they played, and how to best sustain and offer them.

With a data story in hand, the second phase of the data mural process focused on collaborative visual design. We facilitated some shared drawing and ideation activities to collectively create an illustrated symbolic language to represent the story, in addition to the data behind it. Emily synthesized a full draft design based on elements drawn by numerous participants, distributing ownership of the final visual while also allowing for a consistent aesthetic look that holds together as an artistic production. The final C4HW design depicts two young children on a city street lined with supports and resources, all under a sky of clouds. The clouds symbolize, and are labeled with, some of the primary challenges families faced related to health. The street buildings were all depictions of the real locations of local services that could help.

The third phase of the data mural process is the collaborative painting at the site. Like the *How We Fish* gathering, we welcomed a diverse group of participants to help create the final product (Figure 2.9). The invitation to paint opens new doors to working with community data. The C4HW mural was painted in the entrance lobby of the local Head Start facility, where parents often come in with their children to enroll and access services.

The process, outcome, and resulting artifact all opened doors for challenging and constructive dialogs with families and staff about the issues impacting their health. Roughly a year after painting the mural, we were invited back to the site because they were doing some remodeling that required touching

Figure 2.9 A team of volunteers working on a C4HW data mural.
Credit: author.

up and extending one side of the mural. Speaking with the receptionist, who was a talented artist and had done all the cartooning of people in the mural, we found that he'd been using the mural itself as an explanatory tool. In parallel to providing visitors with formal pamphlets describing services and how to access them, he would ask them about their challenges, and then point at the related parts of the mural to refer them to nearby buildings where they should go next to get assistance. He was overjoyed at the explanatory power it gave him, and how parents reacted to the piece. The data story was being actively told and retold, connecting with new audiences over time.

This mural design encodes many layers of data and context, indelibly tied to the specific place it is painted. It depicts knowledge constructed by participatory data collection activities from the coalition, and community experiences only known to those who live there and interact with the services shown. This mural, at first glance, looks like a simple depiction of children in a friendly community; a quick look suggests youth on a supportive path of growth. Looking more deeply one begins to see elements that add detail and nuance to this story, beginning perhaps with the clouds labeled "expensive food & housing," "too little money," and "not enough time." Continued reading reveals relatable questions in talk bubbles of multiple languages, and a cast of characters that reflects the diversity of the community itself. This is echoed in the trail of flags flying behind the child's bicycle, which are based on the immigrant populations in town, as are the subtle greetings of welcome on the road lane dividers. For those who know the area well, another deep look reveals that many of the buildings themselves match the designs of real ones around town where various services are offered, specifically responding to some of the questions in the talk bubbles and clouds. Another insider element is that some of the characters in front of the buildings and in the windows are actual well-known staff who have worked there for years, cartoon versions of recognizable neighborhood personalities.

We've iterated on this multi-stage process for creating community data murals with various groups, and it has inspired work across the world, from Brazil, to Norway, to Kenya. Our assessment activities continue to show that beyond leveraging the power of murals described earlier, we also are increasing data literacy, resetting negative opinions of data, creating and reinforcing community networks, and showing increased willingness to use data to support work amongst participants.[52] The data murals process builds critical data literacy through its participatory engagement with community members.

80 FOCUS ON PARTICIPATION

Learning from graffiti

After our initial few data murals, we pushed for new collaborations and approaches to build our own participatory practice, leading to work with pioneers of the protest graffiti movement in Kenya. In 2012 and 2013 Kenya was in the grips of the political upheaval of a presidential election. To radically simplify and summarize the situation, one could describe it as politicians jockeying for power amidst a backdrop of a population frustrated by corruption and mismanagement. Amid the protests, a burgeoning graffiti movement took up the public calls for accountability and responsible leadership.

Many of these were created by three Kenyan graffiti artists—Uhuru B, Bankslave, and Swift. Their art helped draw attention to the allegations of corruption and more, serving as both political cartoons and public messaging, all through the power of street graffiti. These three defined much of that era's politicized street graffiti and mainstreamed it in Kenya. Uhuru B describes how "Graffiti does address social issues, looking back from the past to now. We have really grown, and we've tackled taboo issues such as corruption, injustice."[53] Their art practice was intentionally controversial, putting a public image to the questions and anger felt by so many around them during the election time.

Built on inspirations from our shared colleague Sasha Kinney, I had the opportunity to work with these three artists to explore the idea of what data-inspired graffiti might look like. How could data be useful for them as a tool to bring people together to inform their process and product? Their messages were already meeting people where they were on the street and supported layered stories via jokes and double meanings. Curious to explore this together and supported by a grant from the Making All Voices Count initiative, we met in Berlin for a series of collaborative sessions to engage and extend both their practice and our own approach to making data murals.

I decided to center on the topic of education and employment data from the Nairobi city budget, and a governmental evaluation of learning outcomes. Following an abbreviated version of the data mural design process the group quickly centered around three key terms they pulled out from reviewing the data: "education," "finance," and "rural." Together in a drab German conference room we sketched out related terms and symbolic ways to represent them.

From these terms and images representing them we continued by developing a zine-inspired data storybook to build a narrative, sketched a symbolic language together, and finished with a draft design of a mural that could be painted. The design told a story of "tipping the scales" that pointed at

Figure 2.10 The sketched out mural design collectively created with Kenyan graffiti artists Uhuru B, Bankslave, and Swift.
Credit: author.

how administration is far more funded than education (Figure 2.10). It used symbols, text, and numbers to satirize the powerful leaders and ask why learning wasn't more well-funded in the budget. While this idea was never painted, the process of designing it impacted all of our work. I walked away with a deeper understanding and renewed commitment to honor and engage the grounded work of community partners. Uhuru B, Bankslave, and Swift spoke of their excitement about the "structure and approach in general as a session," ready to return home and apply their experiences to their own practice as well as training programs they were creating for the next generation of up-and-coming street artists in Nairobi. In a shared reflection afterward, Swift commented that "this approach is fun and should be used all the time." Engaged by the idea, he later painted one in support of a political candidate that included a word cloud showing top terms used in his speeches.[54]

More recently, design scholar Dietmar Offenhuber has theorized another type of public information display he describes as *autographic visualization*.[55] These are made of the physical traces and indicators left behind, or created, as data. In public space an evocative example is the accumulation of pollution on concrete walls around highways. Offenhuber washed this away strategically to create his 2018 piece Staubmarke/Dust Zone (Figure 2.11). His goal was to provoke speculation and conversation about the impacts of air pollution in the world around us (specifically at the site in Stuttgart, Germany).

Figure 2.11 Offenhuber's Dust Zone (Staubmarke), 2018.
Credit: Dietmar Offenhuber.

A 2015 paper from Microsoft research labels part of this power of physical location as "data in place"—the idea that putting data displays in the context and physical location they are about lends them more power and impact.[56] Their concept is one of situated data displays, created to engage with people and places.

Data murals across the globe

As with data work overall, there was a burst of data mural activity driven by the pandemic. One set of evocative examples came from the informal settlements in and around Nairobi, Kenya. The government-imposed curfews and the resulting police enforcement created significant trauma in the area. To document potential human rights violations, the local Social Justice Centres Working Group created and ran a survey, showing thousands of injuries and dozens of killings.[57] In parallel, like in so many other communities, there were strong and sustained actions of resilience and collaborative support.

As pandemic restrictions abated, the group decided to hold a public event to reflect on relationships to authorities. They were inspired to include data as a central element after a training two members received from Data4Change, a London-based nonprofit that supports CSOs working on data campaigns. This led to three painted data murals at the headquarters in Dandora, an informal settlement constructed with World Bank funds near the primary dump for Nairobi. The first documented the key facts and figures from the survey—injuries, arrests, and killings—in stark black and white. The second and third were more participatory, colorful, and offer inspirational examples of creating participatory data murals as a mirror for community perceptions and sentiments.

The second mural began with simple labels for key groups that helped during the pandemic time: health workers, teachers, nonprofits, neighbors, and police officers (Figure 2.12). Working with heart shaped stencils, they invited participants at a festival to paint a heart under a group's label if they felt thankful to them. It offered a participatory mirror to the co-constructed data. This stands in stark contrast to standard surveys, and significantly changes the perceptions of ownership and accuracy.

For the third mural the team created a set of questions about people's identity and perceptions of safety (Figure 2.13). After filling out a short survey on paper with answers, each person was given a string to tie around pins indicating their answers on the wall labeled with questions. The color of their rope represented their gender identity. The final piece incorporated strings from more than eighty people, looking like a brought-to-life parallel coordinates

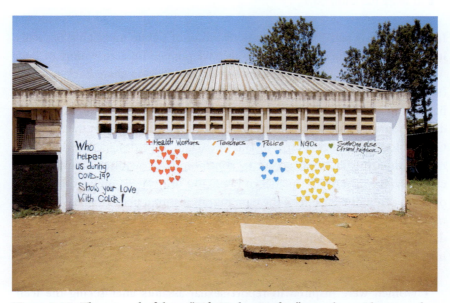

Figure 2.12 The second of three "Life Under Curfew" murals; residents used stencils to paint the responses on the mural itself.

Credit: Social Justice Centers.

84 FOCUS ON PARTICIPATION

Figure 2.13 The third "Life Under Curfew" mural involved tying strings to responses on a wall.
Credit: Social Justice Centers.

chart. The piece went on to win a gold award at the 2022 Information is Beautiful Awards. A team member powerfully described their key takeaway of the result:

> instead of people publishing books that community members are not able to access or read, we put these books on walls, and people can now read them and share the knowledge.[58]

Inspired by this work, and the long tradition of public messaging on walls in that region of the work, data murals have popped up across more of East Africa. In 2019 Data Zetu, the same group that led the share-back sessions (see "Participatory urban planning" in this chapter), created a series of public health data murals.[59] These featured a set of priorities that surfaced from data produced by surveying local populations. Partnering with the Wachata Crew, a local graffiti collective in Dar es Salaam, they designed and painted murals depicting data on access to contraceptives, acquired immunodeficiency syndrome (AIDS) incidence rates, and drug abuse prevention. Their opening ceremony for one of the data murals featured performances by local musicians and was covered by local television, increasing the impact of the data-informed visual. Their follow-up assessment work revealed some striking impacts.

- Women were inspired to report sexual assaults after seeing the teen pregnancy prevention mural.
- Students at a local school referenced data from the murals during a debate.

- Viewers showed a significant increase in their belief that data is relevant and important to day-to-day goals and decisions.

Data Zetu note that this work was specifically inspired by our data mural work, and I'm heartened and proud to see the impacts expanding globally to areas I couldn't have imagined when we first began exploring the idea.

Build participatory data processes

Aligning data storytelling with pro-social sector goals requires you to focus on participation. This can look many ways, from sitting around a table analyzing data, to completing charts online, to painting on a giant wall. Journalists, artists, and activists are constantly innovating new approaches to inviting participation.

A final evocative participatory example brings me back to Boston. Like many American cities, Boston continues to be heavily racially segregated. This is a shameful enduring legacy of historical racist zoning practices called *redlining*—the process of drawing maps to denote where federal loans could be given to support buying a home. Areas of the maps surrounded with red lines were not open to mortgage applications and were almost universally drawn around parts of cities where more Black people already lived. You can still see the manifestations of that policy in modern-day demographic maps.

In 2015 local housing action group City Life/Vida Urbana decided to showcase that continued impact by heading back to the streets, they literally drew the red line down the sidewalk on a public tour. The piece used the power of performance art and public engagement, letting residents physically push a cart that left a red line in its wake on the pavement (Figure 2.14). Even more impactfully, the chalk mark it painted was created by volunteers crushing bricks from locally demolished buildings.

The walk led dozens of residents across the neighborhoods of Roxbury and Dorchester. Looking to either side of the line they could still see the impacts of the mortgage program today. In all too many locations to one side lay large homes with luscious yards, to the other dense decaying housing with little greenspace. Local news quoted participant Marshal Cooper sharing that "it happened in the 30s and it's happening now."[60] A coordinator of the project shared that "[w]e want to inscribe that history onto the streets of Boston, to make that history visible, to make public policy visible."[61]

Figure 2.14 Resident Marshal Cooper paints a red line marking the historical zoning barriers.
Credit: courtesy JamaicaPlainNews.com.

The City Life/Vida Urbana redlining remembrance offers us a final example of activists creatively focusing on participation to project data back onto its location. Walking around the neighborhoods, pushing the cart, hearing the squeak of its wheels, seeing the red trail it left; the entire data experience set participants up to critically reflect about the impacts of the redlining location data on their community. Justice was the goal of the spectacle; data was its fuel.

The dominant toolbox we have for working with data doesn't focus on participation, leading to missed opportunities and coming in direct conflict with standard approaches and methods central to the pro-social sector. A variety of theoretical lenses help us rethink data in society and put it in alignment with goals of empowerment, participation, and social justice. Data for good, critical data literacy, data justice, and data feminism all connect to and inform our data mural practice and many of the other examples shared in this chapter. The key to it all is participation.

3

Build Mirrors, not Windows

Many harms of datafication stem from the concept of data as a window used to see other groups and make decisions about them without any engagement. Data literacy programs have spread, but often focus on this approach. Using data as a mirror instead aligns more closely with goals of pro-social spaces. The practice of creating data sculptures offers us a technique that reflects data back to a community in understandable forms, opening opportunities to engage diverse populations around data about their own community.

Racist data creation in the US

To the unfamiliar (i.e., me), it looks as though US women's clothing sizes follow a logical scale, labeled with even numbers from zero to the twenties. However, that understanding, as any American woman will tell you, is completely wrong. Unlike men's clothing, the sizes aren't trying to represent a real measurement; they are simply numbers naming arbitrary sizes that range from small to large. There is absolutely no consistency across brands, making clothing shopping a fraught mystery hunt full of frustration. The story of how these sizes came to be is an illuminating and surprising look into the racist history of extractive data production, the surprising reach of its impacts, and how disempowering it is to treat data like a window through which to observe others and make decisions that impact their lives.

Our current round of data hype feels new and novel but has strong parallels to the first few decades of the twentieth century. That round of datafication was driven by the eugenics movement, which used scientific processes and data production to demonstrate "proof" for their brand of racism.[1] Massive reams of data were created in service of arguments trying to justify the superiority of populations that at the time counted as white. This had widespread effects that have echoed through time,[2] from the 1960s concept of a "culture of poverty," to forced sterilizations in the US in the 1970s,[3] to ongoing proposals to limit childbirth in developing countries, to modern computer scientists that try to

Community Data. Rahul Bhargava, Oxford University Press. © Rahul Bhargava (2024).
DOI: 10.1093/oso/9780198911630.003.0004

use machine learning and artificial intelligence (AI) to detect individuals that hold threatened identities such as ethnicity[4] and sexual orientation.[5]

How is the data-driven eugenics movement of the 1920s connected to the size of my wife's pants? Journalist Heather Radke dove into the intersection in her wonderfully titled book *Butts, a Backstory*.[6] Her research traces our modern sizes back to a racist dataset created during that era, a dataset that was heavily influenced by the eugenic practices that shaped early statistical methods.

The early 1900s had the United States firmly in the throes of the Industrial Revolution, leading to large investments in the automated manufacture of clothing. This pushed companies toward standard sizes, so they could more easily and cheaply manufacture large quantities. Those efforts began without any standardization. The parallel spread of catalog-based shopping meant significant numbers of women were ordering clothes without any real sense of whether they would fit. Unsurprisingly this led to large numbers of returned orders, a problem the newly established industry Mail Order Association of America decided to work on.

A response emerged from an unlikely place, the desks of Ruth O'Brien and William Shelton, statisticians in the US government's Home Economics Department. The pair proposed to survey women across the country to create a reference set of body sizes; a dataset the clothing industry could then take and turn into patterns that they had evidence would fit more women correctly. Building on capacity in the New Deal Work Projects Administration of the time, they hired one hundred women to lead "measuring squads." These data producers hosted parties across the nation, taking fifty-nine different measurements of each woman who attended (Figure 3.1). This could have been a great approach to creating participatory data-driven standards, but there was one large caveat: while O'Brien and Shelton instructed the squads to invite and measure any women who were willing, they also told them to remove all data from non-white women. At the time this meant removing Black, Italian, Eastern European, and Jewish women. Their data production process was racist at its core.

Women, like men, come in all shapes and sizes. The body shapes of people from around the world vary. Omitting large populations means this hypothetically comprehensive reference dataset was predetermined to serve the sizing needs of the dominant white majority more than any of the other groups. These statisticians, observing the target population from their desks in Washington DC, designed a process to extract data from the women and give it to industry. There was no compensation for subjects, no discussion of what their

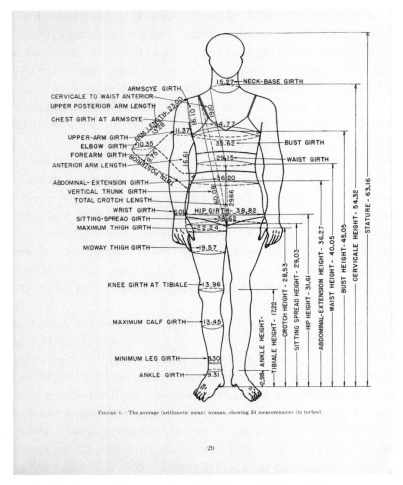

Figure 3.1 The "average" woman from O'Brien and Shelton's dataset, as published by the United States Department of Agriculture (USDA) in 1941.
Credit: USDA, Public domain, via Internet Archive.

needs or interests related to clothing might be, no debate of whose measurements would be included, and no attempts to engage them in the formalizing of the dataset as a reference for future standards. This was using data as a window.

How did we get from this dataset to a size four dress meaning little to nothing? The government published a standard based on this data.[7] Industry iterated on that base over the ensuing decades, and in the end, we ended up with about half of the US population left without any useful fit guidance when trying to buy clothing. The ripple effects continue today; women's inconsistent clothing sizes play a contributory role in developing a harmful sense of self, not to mention the day-to-day struggle to find clothing that fits without useful size information on the label.

The data-driven history of women's clothing sizes is an example of using data as a window, not a mirror. In the pro-social sector, we must intentionally work to avoid this default norm of data production. Data used as a window builds on a position of power in the observer. Data used as a mirror is the opposite, intentionally creating ways to let a community see itself in another way via data. Popular data rests on the goal to *build data mirrors, not windows*. Our recent round of datafication is heavily tied to using data as windows, and it underlies many data literacy programs that have been developed in response to datafication. Using more creative materials than charts and graphs for data representation offers an opportunity to create more data mirrors, like appropriate and impactful data sculptures that reflect data back to a community.

Datafication creates windows

We're in a world where the UN's Under Secretary General wrote a poem about data governance and AI,[8] where data science has become so popular that the Harvard Business Review called it the "sexiest job of the 21st century,"[9] and where I see ads to learn RStudio data analysis software on the back of an auto-rik on the street in India (Figure 3.2). Data related skills are now widely discussed as a new core competency necessary to promote across civil society and business.

Like most technology-associated terms that bleed into the popular press, one could argue late into the night about who originated the term datafication for describing these changes. Most point to a 2013 book by Viktor Mayer-Schönberger and Kenneth Cukier as one of the main amplifiers of the term's adoption.[10] Their book posits that quantification and digitization have created a system of *datafication*, where the world can be seen as a source of data ready for analysis via computational means for insights, predictions, and more. I further specify the definition to refer to systems that have become computational in ways that create quantitative data trails as a by-product of human action and interaction. The datafied society is a construct of our human practices, not some predetermined reality made of data hanging from digital trees just waiting to be plucked.

Figure 3.2 An auto rikshaw in India advertising data science education.
Credit: author.

Datafication has taken hold at the individual, community, and state level. Global governance infrastructures are increasingly using data to measure development and other indicators. Ratified in 2015 by the United Nations (UN), the Sustainable Development Goals (SDGs) are a set of seventeen development and action goals intended to be achieved by 2030—no poverty, zero hunger, climate action, and others. To assess progress, each "goal" is broken down into ten or so "targets," each including a set of quantitative data "indicators" that can be tracked. As an example, consider Goal 2: zero hunger, which includes Target 2.1: universal access to safe and nutritious food, designed to be measured in one way by Indicator 2.1.1: prevalence of undernourishment. A rigorous set of metadata standards accompany each indicator, overseen by a custodial agency (the Food and Agriculture Organization in the example of Indicator 2.1.1). This system of goals, targets, indicators, and data custodians has become the central data framework for measuring global development. The SDGs offer the promise of a data dashboard for the planet.

The process of datafication has created opportunities and economic benefit, but also personal, interpersonal, and structural harms. While the benefits are mostly accrued by dominant groups, the harms have been disproportionately burdened on marginalized groups. We can look at digitization efforts in India for a concrete example of these harms. In 2001 the southern state of Karnataka created the computerized Bhoomi system to systematize and digitize their land ownership data.[11] The goals were to streamline government operations and simplify resolution of land claims. Unfortunately, by creating a central data store of land ownership the state made it easier for predatory planners familiar with digital government protocols to seize large swathes of land and sell them off to developers. This data system directly led to loss of livelihood and land seizures from small landowners; many of whom didn't have the official paperwork the digitized process required despite living for generations on the land. The land dataset was weaponized against the people whose possessions it purported to record. The story of Bhoomi is a cautionary tale about digitized public data and its potential uses as a weapon of power. Systems like this are rolling out across the world; a 2023 study found 113 more advanced automated decision-making systems being used in the public sector in Colombia.[12]

Data illiteracy amongst small landowners in Karnataka made it easier for those with power to steal land via the Bhoomi system. To a certain extent this has always been true of bureaucracies; those who understood the paperwork and processes, and where/how to "grease the wheels" appropriately, were always predisposed to receive more benefits. However, the datafication vastly increases the capacity and reach of those who speak the language of data within the state, significantly exacerbating disparities of access and service.

Build mirrors instead

Asim is late for work. His mother is sick. He spent the night swapping water-soaked cloths onto her forehead to keep her fever down. Now it's time to

94　BUILD MIRRORS, NOT WINDOWS

get to work, where he knows he'll have to face any consequences for being late. He quickly pulls aside the curtains hanging in front of the windows in their small house and walks his sister through the day's care routine. As he runs out of the door the smells, sounds, and sights of the street assault his senses, still groggy from a night of fitful care provision for his mother. He dodges carts, animals, and vendors all preparing for the early morning rush of the regular market. They don't distract him from his worries, both for his mother and about being late. Approaching the factory where he works, Asim rushes to the main entrance as the heat of the morning starts to bead sweat on his brow. His supervisor is outside the gate, making notes on his tablet. Seeing Asim, his brow furrows as he asks for an explanation for why he is late. The manager dutifully marks down Asim's response, logging the infraction into the employee tracking system as his third recent tardy arrival. Too many data entries of late arrivals like this, no matter the reason, and Asim is likely to lose his job. Head down, he walks into the factory to start his workday, thoughts still on his mother's condition.

Anyone that has worked in the service or industrial sectors has had mornings like Asim. But his story isn't from our time. Asim's late arrival wasn't entered into a digital time tracking system via an iPad; his tardiness is recorded for history on a clay tablet from over 3,000 years ago. This tablet is known as an *ostracon*—a broken piece of pottery re-used as a writing surface—and dates to Egypt during the reign of Ramses II in 1250 BCE (Figure 3.3). This tablet records people clocking in to work as a column of names and dates written in black ink. Above each, in red, is a reason for any absence by the individual on that day. These include explanations such as "daughter was bleeding," "brewing beer," and "wrapping [the corpse of] his mother." We can only assume this was to dole out some punishment if the workers were absent too often. We'll never know what happened to Asim, but he'd surely be surprised that his tardy arrival is noted on this permanent record stored at the British Museum thousands of years later.

The history of data production, which created and informed our current practice, is enmeshed in stories like Asim's: an individual or group with power of control over another group uses data to track and influence them. These are the data windows I've been describing so far in this chapter, and they dehumanize and separate through the often-convenient trick of quantification. However, data production through a window limits our view, making it hard to see multiple perspectives of any phenomenon or setting; the data producer can't see past the bounds of the window.

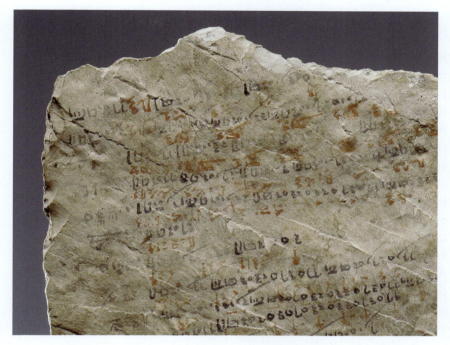

Figure 3.3 Closeup of the ostracon EA5634, recording reasons for being late to work.
Credit: The British Museum.

Popular data proposes an alternate approach: we should use data more often as a mirror, not a window. We use mirrors every day to validate our self-perceptions. We use mirrors to compare ourselves to others. We can even use mirrors to look around corners, predicting what might come our way next. Mirrors offer a more appropriate and useful metaphor for thinking about ways to engage people around data in alignment with goals for community engagement and empowerment.

Data as a window externalizes the control of impacts from data production and analysis to outside a community. Data as a mirror gives us agency to act on data that records our actions.

Data as a window offers insights from a third-party perspective. Data as a mirror lets us learn something about ourselves.

Data as a window allows us to blame the observed other for undesired behaviors depicted in the data. Data as a mirror forces us to acknowledge that the good and bad we see in the reflection is us.

Build data mirrors, not windows.

Interactive displays as mirrors

Fast forward a few thousand years from Asim's story and we can explore how modern technologies facilitate digital data reflections that are interactive in public space. It is now common to use web and mobile-based live survey tools that make audience participation easy and fun. I see these being used everywhere from my children's elementary school to conferences I attend. That's the simplest example of a public digital data mirror, and computer human interaction researchers have worked to tease out more detail about how they work.

One set of evocative examples is a series of projection-based interactive voting projects from researcher Anna Valkanova. In 2013 and 2014 they iterated on approaches to public interactive data visualizations to study the participation and deliberation they provoked. The first, REVEAL-IT!, solicited monthly energy expenditures and displayed results in cultural centers in Córdoba, Argentina and Barcelona, Spain (Figure 3.4).[13] The team found that the projected data reflection of energy costs functioned as an "urban communication tool," motivating questions about individuals' role in overall energy consumption. Participants in the study also spontaneously brainstormed

Figure 3.4 Researchers speaking with community members in front of the REVEAL-IT! visualization of monthly energy expenditures in Barcelona.
Credit: Courtesy Nina Valkanova, Sergi Jorda, Martin Tomitsch, and Andrew Vande Moere.

about amplifying the data and their perceptions of it though other mass media to impact community-level perceptions and actions around conserving energy. Looking into the mirror showing their energy data provoked social communication between community stakeholders about issues of civic import, much like the public street murals of energy consumption I shared earlier from the Tidy Street Project (see "Painting Data on Walls" in Chapter 2).

A follow-up project, named MyPosition, explored the power, potential, and perception of playful interactive public data visualizations.[14] The team here focused on deliberation by creating an interactive embodied voting system; viewers could move their bodies to cast a vote in front of a projected screen and see various summaries of aggregate results (Figure 3.5). A prompt on screen asked about agreement with the statement that a "bachelor's degree is the best preparation for a job," motivated by the installation site at a large university. Through various iterations of the interaction and display design, the team found that the project sparked engaged civic interaction and discussion. The interactive mirror increased participation, a central goal of many pro-social settings of data use. Their playful invitation to interact via body positioning also led participants to push one another to participate, perhaps building on viral or pro-social peer pressure dynamics of public settings.

This series of interactive projected displays point at growing understanding from research about how data visualizations can work as public mirrors

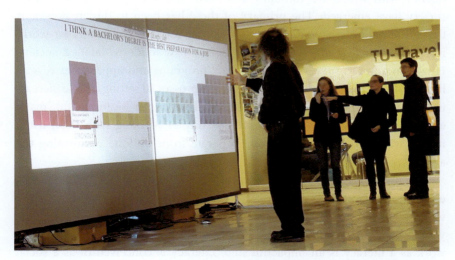

Figure 3.5 A university community member moving their hand to vote in front of the MyPosition piece.

Credit: Courtesy Nina Valkanova, Robert Walter, Andrew Vande Moere, and Jörg Müller.

to encourage participation and deliberation. These were short-term academic projects, but still suggest an underlying approach that can be built on in real-world community settings. Returning to inspirations from the world of community murals (see "Paint Data Murals" in Chapter 2), artist Candy Chang's *Before I Die* series of murals function in a similar way. She paints a large wall in all black and covers it in a stenciled grid of white text prompts all reading "Before I die I want to _____;" the public is invited to fill in the blank with chalk.[15] The commitments and dreams shared on the walls are powerful, poignant, and sometimes peculiar. She began the project in 2011 and since then it has been replicated across the world by others on over 5,000 walls in eighty countries. Participatory data mirrors, whether high tech or low tech, provoke participation and reflection.

The concept of data literacy

I offer windows and mirrors as useful metaphors for thinking about how data is used, by whom, and for whose purposes. The concept of data mirrors pushes us to create opportunities for broader populations to have a seat at the table in datafied community settings. The growth of datafication across individual, community, and state levels is increasing the frequency with which we run into this need, and multiple responses from society have emerged. The primary of these has been to design programs that build *data literacy*—a concept that posits a basic set of skills for working with data that are required for living in a modern datafied society. However, a broad survey of data literacy programs and approaches reveals limitations for justice-focused projects in the pro-social setting. While increasing individual statistical and computational skills can be impactful, many efforts are too narrowly focused on one way of working with data, using it as a window to analyze a group separate from the one performing the analysis.

Data literacy is driven by the growing use of data in business, culture, government, and society writ large. The simple act of existing and interacting with digital systems produces large amounts of data stored across the globe, yet working with this data is a skill only some have access to. Naming data work as a literacy is an impactful step. Literacy is conceived of as a social good, supporting community-building, participation, communication, learning, and economic development. Yet what does data literacy entail? Is it truly a critical new set of skills?

THE CONCEPT OF DATA LITERACY   99

Governments are acting to build broad data capacities, driven by perceived educational and labor needs. As just one example, in 2023 the US House of Representatives took up a bill, the Data Science and Literacy Act, that would direct the Department of Education to create a new grant program specifically in support of building data literacy skills.[16] The bill argues that "data literacy is essential for both effective citizenship and well-being," and goes on to link it to workforce development, public communication, and media consumption. This is just one of several major efforts to support and educate the larger populace on analytic skills connected to data, from "pre-kindergarten through postsecondary levels" (as directed in that bill). Parallel efforts are outlining and fleshing out data literacy skillsets as curricular standards that would encourage widespread adoption across public schooling in the US.[17]

A single shared definition of data literacy is contested; the fields of statistics, computer science, and mathematics all claims ownership of approaches to working with data and define associated literacy in different ways. *Numerical literacy* relates to mathematical operations built on quantitative abstractions represented as numbers. *Information literacy* is about understanding information sources in ways that help drive critical thinking. *Scientific literacy* narrows in on the formal definition of the scientific process and how it can be put into use to aid in understanding the world around us. *Statistical literacy* connects to modeling relationships and analysis of numbers in support of navigating the world. *Computational literacy* builds on the basics of algorithmic thinking and modeling as an approach to problem solving using computational technologies. *Digital literacy* pertains to using computational tools, with the internet, to access services, perform tasks, and more. Each of these literacies informs definitions of data literacy. In fact, some definitions propose that what makes data literacy unique is the fact that it operates as a combination of these literacies, living at their intersections like no single other literacy does.

The growing importance and use of data has driven the creation of large numbers of educational programs that build data literacy. These programs have embedded in them answers for the questions of what data literacy entails, and who it is intended for. The majority do not focus on educational approaches that center empowerment and justice; they focus on technical skills acquisition by individuals rather than goals such as empowered engagement that are so central to the pro-social sectors where data is newly being used. This presents a problematic gap and unanswered questions for data literacy efforts: are they just training more people to build computational data windows? Or are they building critical capacities for empowerment in the tradition of Paulo Freire?

100 BUILD MIRRORS, NOT WINDOWS

Literacy as a path to power

Mention "literacy" to someone and it immediately conjures up images of learning to read, with positive associations about critical skills necessary to live in a society. The first few years of public education in the US for children are hyperfocused on learning to read. The UN lists reading literacy as the fourth of its Sustainable Development Goals. Naming a "data literacy" is thus an impactful statement, building on this popular understanding of its importance and positioning the skills of data work as critical for both the individual and society at large. A literacy of data suggests that related skills are necessary for a broad population to acquire, not a specialized skill set reserved for a few.

Some prominent historical Western thinkers have challenged this framing of literacy as a net good, with a specific emphasis on questions of power, agency, and subjugation. In Greco-Roman times philosopher Socrates (as described by Plato) offered that in teaching written language you provide your students with the appearance of wisdom, not with its reality. His argument paints literacy as the privilege of forgetting, a skill that allows for new and less-thorough ways of knowing something, precisely because synthesized knowledge can be externalized from an individual by writing it down.

Twentieth century French anthropologist Levî-Strauss went further than Socrates in his critique of literacy, arguing in his *Tristes Tropiques* that "the primary function of written communication is to facilitate slavery."[18] He makes this startling (to most) claim after asking what advances in society reading literacy has allowed for, responding with a focus on cities and systems of classification, and thereby concluding that literacy programs in modern societies follow "an increase in governmental authority over the citizens."[19] These historical perspectives point at ways to think about how reading literacy is deeply connected to questions of power and access. Specifically, power as it pertains to both self-determination and power as it relates to exercising control over others.

In contrast to those two men's perspectives, the dominant modern understanding of reading literacy is one of empowerment and knowledge building. Literacy to most is a doorway to engaging with society's dominant structures. However, it is important to remember that this door has always been shut to some, based on various aspects of their identity. Women have long been denied access to education, and learning to read, as a means of control and subjugation. Medieval King of Spain Philip III wrote "Women should not learn to read or write unless they are going to be nuns, as much harm comes from such knowledge."[20] In the US, laws such as the nineteenth century Alabama

THE CONCEPT OF DATA LITERACY   101

Slave Code codified the restriction on literacy and imposed a penalty, stating that "any person who shall attempt to teach a free person of color, or slave, to spell, read or write, shall [...] be fined in a sum."[21] The enslaved, recognizing the power to be gained through literacy, routinely fought to find secret ways around these laws.[22] Reading literacy is part of the toolbox for restricting access to structures of power, socially and legally.

These perspectives leave us with questions about the meaning and role of literacy in society. Is the history of reading literacy just one of subjugation and restriction? Or is it a pathway to engaging with power structures that has led to greater social equity? We can return of Brazilian educator Paulo Freire for a model that is more apt for the pro-social settings of current data literacy I focus on here.[23] For Freire a concept of literacy is built within the context of social empowerment and individual transformation. His *educação popular* (popular education) practices were designed to help the large under-educated classes in society, using literacy as a tool of empowerment. His work sought to build the ability to identify, resist, and interrogate systems of inequality (critical consciousness). Rethinking the teacher-student relationship, Freire's approach puts educator and learner in partnership as compatriots whose understanding is broadened through interaction. This is an approach to building literacy that centers goals of justice, social development, and equity. This is an approach to building literacy that aligns with the newer pro-social settings of engagement with data.

The history of literacies leaves us with competing conceptions of what specifically data literacy might mean and what goals it might advance. Do technical skills-based training in data tools further entrench power structures, as Levî-Strauss might argue? Or do introductions to working with data offer a path for engaging otherwise marginalized groups in ownership and governance of their communities, as Freire demonstrates? The only conclusion I draw is "yes," to all these types of questions. In formal or informal learning settings, the specific practices, and mechanisms for introducing core tenants of data literacy might unconsciously bake in outcomes the educator hasn't considered. In our datafied society it is not a stretch to think about working with data as a new core literacy. The devil is in the details, of course. The key is how we conceive of data literacy, and specifically whether you focus on increasing capacity to build windows or mirrors.

One evocative nascent example of putting this into practice comes from the Massachusetts Institute of Technology (MIT) Liberatory Computing project. Graduate student Raechel Walker's 2023 thesis documents a high school data literacy curriculum specifically focused on fighting systemic oppression with data.[24] Walker centers around a concept of data activism defined as "utilizing data, computing, and art to analyze how power operates in the world, challenge power, and empathize with people who are oppressed."[25] The program, designed for African American computer science students, leads participants through activities designed to enhance racial identity, critical consciousness, collective obligation, liberatory academic identity, and activism skills. Students built data science projects in partnership with local community groups, leaving them with a strong desire to continue to combine data science and social justice in their work. This type of educational work centers justice and empowerment, facilitating students' technical learning about how to build data mirrors in community settings. While the computational focus of data literacy training like Walker's isn't problematic per say, it doesn't target broader populations in the way a literacy like reading does. How do data literacy programs appeal to audiences who aren't predisposed to work with math and on computers?

"I'm not a math person"

In 2014 I had a chance to visit the classroom of someone that greatly influenced my own informal journey into learning data science—Professor Allen Downey. His technical books on programming and statistics are must-reads for anyone learning to do data science with the Python programming language and were particularly valuable for me as I tried to fill in gaps in my own self-taught statistical skills about Bayesian analysis and related topics. That year Prof. Downey generously invited me into his Olin College classroom, just outside of Boston, to introduce his students to a different approach to storytelling with data. Nervous about teaching in front of someone I had learned so much from, I approached the session with some trepidation, worried about coming from too far out of left field with my take on data storytelling. One of the first interactions with Prof. Downey set me at ease, and introduced a way of thinking that has stuck with me ever since.

As I contrasted storytelling with statistical analysis, he jumped in with a helpful way of talking about math. He pointed out much of statistics, at its most basic, is just counting. At its most complex, it can include rigorous statistical methods (represented with lots of Greek letters). There is, Prof. Downey argued, a lot of space in-between.[26] The in-between includes many techniques that offer analytic power to non-statisticians. Aggregation is just adding. Normalization is just dividing. Binning is just rounding. Here was someone that

THE CONCEPT OF DATA LITERACY   103

wrote a widely read book on programming and statistics for data science arguing that you didn't have to be a statistician to produce many types of useful and robust data analysis. I've hung onto this idea ever since, introducing it over and over again in workshops.

Why is this so important? Consider for a minute the idea of your own relationship to topics you struggled to learn. I, for one, don't consider myself a "drawing person." I never developed the visual artist's eye for seeing and creating with pencil on paper. Others might not consider themselves "sports people," feeling uncomfortable moving their bodies in practiced intentional ways. In my data workshops I consistently run into participants that tell me about how they aren't "math people." In the US context I trace that to failures of our public education system, which rigidly introduce a single propositional way of knowing math, systemically pushing away those that struggle with it (there is a history of robust critique of the form and intent of US math education).[27] Ask around your friends and I'm sure you'll find a few who describe themselves this way, as "non-math people." But what does that really mean? Surely they can count, divide, and round. As Prof. Downey told his class and me, you can do a lot of data analysis with that list of simple skills.

Academic studies have validated this observation more formally, finding a common thread of people that define themselves using this am/am-not duality in relation to math identity, more often than with other subjects. Alison Leonard, a Professor of Arts & Creativity at Clemson University, found this phenomenon so befuddling that in 2019 she decided to explore it more in depth.[28] What she found validated her experiences, and my own—people consistently defined themselves as "(non)math." Her study attempted to reset these binary definitions, exploring physical movement-based introductions to geometric transformation with students in a teacher education program. Through that process Prof. Leonard found that people were actually much better at math than they thought they were. She was helping people to engage in new ways with math abilities they didn't think they had. This matches my own experiences with learners. If you engage people in ways that make sense to them, and honor the expertise and approaches they bring, then they can excel at basic math-based data analysis despite any overly pessimistic self-assessment. This binary self-definition related to math ability creates a significant internal barrier when starting to work with data, even in non-math-like settings.

This false binary in self-perception matters because most introductions to data literacy sit in this type of mathematical domain. Programs tend to center

104 BUILD MIRRORS, NOT WINDOWS

technical and statistical approaches to analyzing quantitative data. Browsing a mix of educational[29] and professional training[30] for non-technical domains, one finds statistics and computation are most often core, introduced without significant discussion of context and ethics.[31] Corporate training programs emerged from information technology (IT) departments. Educational programs live in schools of computer science or statistics. This pattern follows a long arc of abstract mathematical thinking being given primacy in American educational curricular guidelines. The current manifestation of this history in the US is the *STEM*—Science, Technology, Engineering, and Math—approach to curricular priorities, proposed as a response to predicted economic workforce needs.[32]

Data literacy has been strongly associated with mathematical literacy, creating a spillover effect where many novices think of themselves as "non-data people" because they already think of themselves as "non-math people." In prosocial settings with goals of engagement and efficacy this creates a significant barrier to entry; another example of how the original domains of data work have created challenges for data use in new settings.

Not only does this technical focus limit who feels comfortable learning to work with data, it also ignores elements of working with data that areas such as math and technology don't help with. My long-time collaborator Catherine D'Ignazio and I decided to tackle this head on while building one of our first data literacy activities. We did it by asking people to think about an unusual dataset: unidentified flying object (UFO) sightings. The activity we designed is called WTFCSV, a reference to a common beginning of many data processes, when one is simple trying to figure out what the f**k (WTF) is in a comma separated values (CSV) file of data that you've just received with no background. The associated web-based activity offers some sample datasets and shows summaries of each column they contain. Our key learning prompt asks participants to brainstorm questions they might want to ask the data based on the aggregate visual representations of each column's values the tool shows (Figure 3.6).

This prompt, about asking good questions, is the type of core data literacy skill that technology doesn't help with. The UFO sightings dataset, one of the samples included on the website, has turned out to be a perfect way to demonstrate this. Workshop participants that choose this dataset, made up of calls to the National UFO Reporting Center about sightings in Massachusetts in the mid-2010s, typically generate questions that look for temporal correlations between sightings and airplane flights or weather events. As people who don't think aliens are flying around overhead in spaceships (perhaps

THE CONCEPT OF DATA LITERACY 105

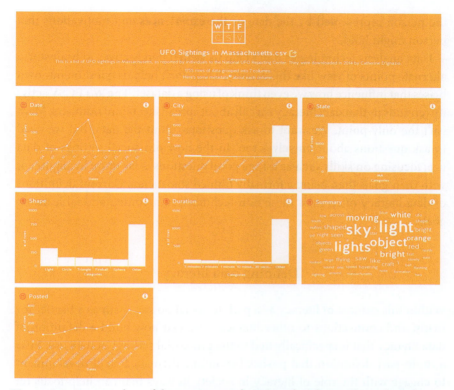

Figure 3.6 A screenshot of the summary of unidentified flying object (UFO) data presented as an example in the WTFCSV activity on the DataBasic.io website.
Credit: author.

kidnapping people and cows now and again), these learners are looking to explain away the sightings as misunderstandings of well-known natural and human phenomena.

In my role as a facilitator, I push back, reminding participants that this dataset was logged from people actually calling to report something they didn't understand, something that they suspected could be aliens flying around. The caller might have a different set of questions based on this set of beliefs. They, in contrast to most workshop participants, might wonder what location would be the best place to see a UFO? What time of year are sightings most common? Can we build a predictive model to help us identify potential places to make first contact? These questions aren't ones participants of the WTFCSV activity typically come up with, because they share a dominant societal perspective that aliens aren't visiting us. Participants don't consider the point of view of the person represented by the data or the experiences and motivations that person might hold.

Data analysis technologies don't help us switch perspectives to brainstorm alternate points of view like this activity. They don't help us question our own biases that influence how we even think of using a dataset. The WTFCSV activity reminds us that data literacy isn't all about spreadsheets and numbers. Math isn't the only point. We want to ask questions about the data, but we need to ask questions about ourselves too. In the new pro-social settings of data use, focusing on skills such as asking good questions is far more in alignment with their goals than being a formally trained statistician. Current definitions of data literacy often miss these non-technical, non-computational learning goals.

Defining data literacy

Within this context of literacy as a path to social power, math as a barrier for many, and connections to other literacies, we can now build a definition of data literacy that is specifically tied to the pro-social settings of data use. I use a multi-part definition that pushes beyond mathematical skill development to engage with the role of literacy in society in ways that can help focus on using data as mirrors. Data literacy must include considering impacts, current uses of data, how to understand and critique it, approaches to representing data stories, and applying those to the world. My definition is action oriented, focused on using data for some pro-social purpose. I break data literacy down into five core competencies.[33]

1. Understanding the process and mechanisms of data creation and intended use.
2. Creating data and assessing reliability of sources.
3. Analyzing data via validated and appropriate methods.
4. Representing data in accurate and meaningful ways.
5. Constructing reasoned arguments with data within some social context.

The first component focuses on understanding the ongoing roles data plays in society. Data is produced by some process, person, or organization for some reason, implicitly or explicitly. The process and motivation of that data creation is critical to question and understand as part of engaging with any dataset. In an era of Big Data we have trails of information left behind us across our digital devices, whether we know it or not. In the US most of this is driven by capitalist market incentives; this data trail is used to build a profile of individuals and groups to be monetized via reselling across industries. These mechanisms of digital data production and capture exist all around us.

Producing, collating, and considering data is the second piece. This focuses on the ability to analyze data already created. There is an overlap here with digital literacy, in that much of this work happens in online settings using computational means of discovery (Google, Excel, Tableau, etc.). Additionally, this addresses how to generate data that is otherwise missing. This competency also overlaps with statistical and scientific literacy, where an individual is questioning and understanding whether a dataset is accurate and can be useful for an inquiry—or hypothesis-driven process.

The third component is the process of analyzing data. This specifically relates to understanding how to perform appropriate simple or complex analytic operations on qualitative or quantitative data. Smaller sets of data lend themselves to observation, manual labeling, or basic numerical analysis. Larger sets of data are usually best approached with computational resources for digitization and analysis in mind or split into smaller sets for collective human-based analysis. As algorithmic techniques become more commoditized, they have moved into software-based black boxes, creating a barrier to accessing and deploying them, no less learning how they work. This, in turn, has driven introductions to data analysis to be built around up-skilling on technical tools, deeply connected to statistical literacy. Due to historical norms, this creates attention toward quantitative data analysis; qualitative approaches are less often engaged and more poorly supported in computational tools.

Representing data is the fourth component and the focus of this book. We transform data into a variety of representations intended to build understanding in a diversity of settings. These depictions often come together in the form of a story, stitching together data with some narrative arc designed for a specific audience. This overlaps with approaches to visual literacy and understandings from cognitive science of how we perceive and process information. Our standard toolbox of representations has focused on the visual at the expense of our other senses (see Chapter 5). The new settings of data use require engaging a broader set of perceptual approaches to reach audiences in effective ways with goals of engagement and empowerment in mind.

The fifth and final component I include returns to how data operates in society; we need to be able to use data to make meaningful arguments within a socio-political context. This synthesizes all the other competencies; understanding the data we have in hand and analyzing it to represent it allows us to use it as fodder for an argument or claim. This is a critical distinguishing trait of data literacy for me and connects directly to pro-social community settings and institutions. This is a literacy that is meant to be operationalized in service of social change due to the social impacts of datafication.

If data skills are a new modern literacy, then it functions as a doorway to engaging with society in productive and empowering ways. The key to this door is then centered on learning the language of data. Defining what this new literacy entails is necessary to understand how to unlock and open this door to as many as possible, eliminating various types of barriers to learners. The definition one chooses implies who data literacy is for, and whether it is used as a mirror or a window.

Who is data literacy for?

The varying definitions of data literacy have spread with the embedded assumptions of the fields that they emerged from. In business settings these efforts have centered on tool application, driven by the ownership of data programs within information technology departments. Questions of power, ownership, justice, access, and more still lurk as shadows in this work, and need to be addressed directly to fit data literacy programs to the pro-social settings it has moved into.

Within the specific settings of data in public spaces and interactions, the competency related to making arguments with data takes high priority for me amongst the others. As data science was introduced and claimed by the field of statistics, quantitative analytical approaches became the standard at the expense of creating a more inclusive set of pathways into the field of working with data.[34] Any introduction to data framed through a mathematics lens immediately raises a massive barrier to the group identifying as "non-math" people.

The growth of data literacy programs has still left large sets of those without data skills excluded from data centered decision making processes. A 2020 report from Accenture and Qlik found that just 21% of the global workforce was confident in their data literacy skills.[35] From personal experience I'd say that number in the non-professionalized class is far lower. A typical room of participants in one of my workshops reacts to data with a feeling of fear and uncertainty; they think data is numbers and numbers are scary. I've seen this repeatedly in my workshops, to the point where many of the activities I've designed have "reduction of fear" as a primary learning goal. Popular data requires an acknowledgment of that on-the-ground reality and must engage efforts to honor and overcome it. I've seen data literacy as a field that engages this fear barrier on two fronts—math and technology.

Companies and educators have launched numerous efforts to address these barriers for the pro-social sector I'm focused on. Google has sponsored the creation of playful tools to create data visuals, trying to support novel approaches to data representation.[36] The Knight Center at the University of Texas and Investigative Reporters and Editors Inc. (IRE) both run an extensive schedule of conferences, in-person, and virtual training for data journalists. Along with Catherine D'Ignazio and the Emerson Engagement Lab, I offered a virtual training course to librarians on how to build their data literacy skills.[37]

The broader data literacy training focus on creating computational and statistical data windows isn't helping communities work with data in ways that engage people that have many ways of knowing. Data literacy that engages the context and goals of pro-social institutions doesn't have to focus on creating charts and graphs. Data sculptures offer one novel and appropriate tool for community settings, providing new ways to design data mirrors that engage non-math audiences with data processes and representations.

Data sculptures as mirrors

Chicago on July 4th, 2020 was shaping up to be an idyllic US Independence Day. Temperatures that Saturday rose to 85° F; the sun was shining. Water balloons were being filled across the city as eager children prepared for playful battles to come. Family gatherings, back-yard parties, lemonade, watermelon, BBQs—these are popular ways Americans celebrate the holiday. The overall experience is patriotic, but also serves as a simple excuse to enjoy the summertime with the big three Fs common to people across the world: friends, family, and food. For our 4th we even throw in a fourth F: fireworks.

Against this backdrop, Chicago organizers at the Black Youth Project 100 (BYP100) decided to build a new tradition. They took to the streets to protest, honoring their community's history of action in the face of injustice. This was their testament, and challenge, to the story of freedom in the US. Their tool to protest with was data, and they were bringing it to the street in the form of a larger-than-life data sculpture.

In 2020 a massive wave of protests swept across the US in the spring and summer, demanding changes to how police are funded as part of the larger

110 BUILD MIRRORS, NOT WINDOWS

Black Lives Matter movement. The call to "defund the police" emerged as an actionable agenda from the summer of peaceful protests in opposition to structural racism in the US, and specifically to the wave of violent killings and murders of unarmed Black men by police. The movement argued that local funds were overspent on supporting violent armed police forces instead of being distributed to social services that serve as protective factors against crime and violence. Debate over the "defund" phrase itself unfortunately eclipsed popularly supported recommendations, such as proposals to divert mental health related 911 emergency calls to non-armed responders.[38] As part of the work activists began deep dives into budgets all over the US, comparing funding rates for police to those of social services, arts, libraries, and more. In general, the results provided strong evidence that reallocating police funds would improve public safety.[39]

Here's the problem for communicating this evidence: budgets are boring. People understand that money matters, and that budgets represent priorities and values, but little tunes audiences out faster than a set of bar charts showing budget allocations for their community. The movement wasn't happening in budget review meetings. The movement was happening on the streets, where peaceful protests were met by a wave of violent police response, leading to even more protests against the forceful suppression of legal public assembly.[40]

That sunny 4th of July in Chicago BYP100 tackled this head on by creating a physical bar chart as part of their protest on the street. The stacked and painted cardboard boxes showed budget allocations for the city. In their words: "These boxes visualize Chicago's city budget. The tall stack is the $1.8 billion police budget. Housing & public health each only have one box."[41] This bar chart, brought to life with the spectacle of street art, provokes reflection and conversation. You simply cannot ignore a giant chart made from cardboard boxes towering over you. Their emancipatory goals resonate strongly with my linking of popular data to Freirean ideals of building a critical consciousness; this example is using data in support of arguments of liberation and freedom from structural violence.

Their street-corner cardboard construction used data as a mirror for the community. BYP100 looked at available budget data to remix a common graphical form with novel materials. The cardboard's simplicity can blind you to the rhetorical complexity it affords. A standard bar chart printed on paper speaks with authority and rigidity; the bars are fixed and accurate. This cardboard chart allows for change; you can pick up a cardboard box and move it. This chart says change is possible, the opposite of a printed bar chart. Their performative act of stacking and restacking spoke to this directly; the budget can be changed. The choice of cardboard as a medium worked in concert with the goals of the organization, reinforcing their message that things can change—it builds efficacy. This example offers a glimpse into some of the power of using the creative technique of physical data sculptures to act as mirrors for a community.

Data sculptures bring information into our world of three-dimensional shape, touch, texture, and manipulation. In traditional art settings, sculptures are meant to be seen from all sides, with a select few built to be touched. The act of representing data in physical form is often described as *physicalization* in academic settings, as a parallel to visualization. Designers of a physicalization might vary the texture or the size or shape in three dimensions based on data. To further specify the terminology in this space, I describe body-based data representations as *embodiment* (discussed later in this book), and object-based ones as sculptures. For this reason, I choose to use the language of data sculptures to describe data representations like BYP100 street chart, because data embodiment offers a wholly separate language with its own richness.[42] A data sculpture is a constructed three-dimensional (3D) representation where physical attributes are varied based on data. With this definition in hand, we can explore why and how data sculptures are a rich addition to the toolbox for storytelling with data in community settings, well-suited to building data mirrors instead of windows.

Haptic visualization

Our sense of touch is a rich channel we use to perceive and interact with the world. Our ability to recollect the feel of an object, called *haptic memory*, is quite strong.[43] Imagine the feel of your favorite sweater, or the quilt you've had since you were young—these physical interactions build evocative sensory memories.

The idea of *haptic visualization*—data representation meant to be understood by touching it—is related to my concept of data sculptures. Looking back again to Indigenous manifestations of data offers three evocative examples (my own research revealed the same three introduced in Hall and Dávila's discussion of "Disruptive Histories").[44] The first is the often-highlighted South American quipu, systems of connected knots on strings used for encoding and transmitting information about food stock, census statistics, economic production, and more (Figure 3.7).[45] These were created and read by touching them and were in use for a few hundred years, beginning around 1200 BCE. A second example of Indigenous haptic visualization are the ocean swell charts created by residents of the Marshall Islands in the 1800s and 1900s. These were

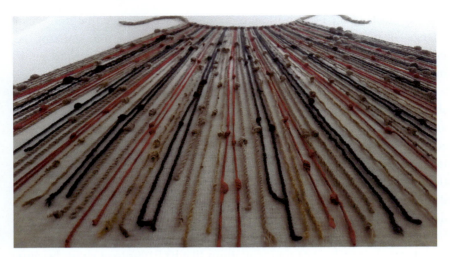

Figure 3.7 An Incan quipu.
Credit: Pi3.124, CC BY-SA 4.0, via Wikimedia Commons.

Figure 3.8 A navigational start of the Marshall Islands.
Credit: Cullen328, CC BY-SA 3.0, via Wikimedia Commons.

intricate networks of sticks tied together to encode the locations of islands (shells) and ocean swells (sticks) (Figure 3.8).[46] Their stick-based navigational charts were produced by a select few experts and read by an expert navigator when taken out to sea in a cluster of canoes, laying down on the bottom of a boat to feel the movement and connect it to the data about ocean swells represented on the map.

A third example comes from the Inuit peoples of the Ammassalik area, in what we now call Greenland. They encoded the geography of the coastline and islands into portable tactile maps carved out of driftwood (Figure 3.9). These highly detailed maps of islands and the coastline were created to be "read" by fingers in gloves, in light or dark, and float in case they were dropped into the sea. The examples we have now in the West were created by an Inuit hunter named Kunit around 1885 and traded to Danish explorer Gustav Holm. Reflecting the colonial thinking of the time, these intricate maps were questioned by other Danes who doubted their accuracy and whether an Indigenous person would have had the skills to create them. This carving told a story of the creator's cartographic expertise in the geographic area, yet European experts immediately doubted it and concocted elaborate tales to explain how the novel technique could have been learned from a ship-wrecked European.

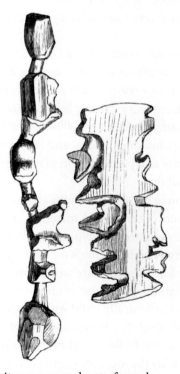

Figure 3.9 Wooden intuit maps carved out of wood.
Credit: Gustav Holm, Vilhelm Garde, Public domain, via Wikimedia Commons.

The concept of haptic visualization reminds us of the opportunity in re-engaging our sense of touch for data representation. Our fingers can read texture and shape in detail and connect to different parts of memory than purely visual displays. Data sculptures can be more than just visual mirrors, they can be haptic ones as well.

Data sculptures as a learning tool

My undergraduate training was in computer engineering, but for my graduate studies I pivoted into education. That pedagogical training has stuck with me, and early in my career as a data literacy educator I started creating activities inspired by my graduate work. Yet, in early 2011, I found myself stuck on creative activity design for a challenging workshop I was scheduled to facilitate. I had been asked to bring together leadership and staff from a network of Boston-area nonprofits to help them build their data literacy skills in support of program goals. Organizers had sensed both excitement and hesitation from the group, particularly related to dynamics around math skills. Leadership was all in on math and spreadsheets, but staff were hesitant and felt disempowered by that reluctance.

Putting on my educator hat, I realized that I needed some kind of ice-breaker activity that would diffuse the tension in the room and reset the power and agency imbalances my collaborators had described to me. I found my inspiration in the world of participatory urban planning, specifically in the work of James Rojas and John Kamp. In the early 2000s the pair built a practice of using craft materials, children's building blocks, and recycled plastic to invite people to reimagine their city and participate in planning activities by physically reconstructing models of their community.[47]

Inspired by their approach, I collected pipe cleaners, wooden blocks, googly eyes, pom-poms, and other playful materials. To spark interest from a wide group of people I designed a one-page handout describing two different datasets about a very relatable topic: ice cream. With these materials and data prompts I ran my fist data sculpture workshop activity, with the goal to break down power dynamics and invite in math-shy participants by using a participatory invitation to physically build with craft materials.

At the workshop I dumped this all out on a set of central tables and gave teams a short seven minutes to build something that represented anything they saw in the data. Very quickly the room took on a positive buzz and energy, with both staff and leadership accepting the invitation to be playful. Leadership in suits took off their coats and literally rolled up their sleeves; data experts worked side-by-side with data novices. At the allotted time I yelled over the din to bring the activity to a close and led a share-back. One group created a line of ice cream cones stacked high with scoops based on the amount of per-capita ice cream eaten per decade. Another simply created a giant bucket and labeled it "chocolate," telling the story of how it was the most popular flavor in the survey data on the handout. People laughed and nodded at one another's sculptures, and suddenly data wasn't something intimidating. They had all moved from data to story in under ten minutes, building collective confidence in their abilities to work together around data.

The data sculpture activity works as a learning tool to engage the participants via a physical invitation to build, breaking down internal barriers and resetting externally perceived power imbalances. In the years since then I've continued to refine and iterate on the activity as I've run it in diverse settings around the globe and online. I found that it works cross-culturally (with appropriate materials) and across diverse age ranges. The activity very specifically uses the power of data sculptures in an introductory setting to overcome social and technical barriers to working with data. Through activities like this one, and longer design exercises in university and professional settings, I've found that making data sculptures is especially well suited for data literacy learners.[48]

In more formal educational settings, we can find statisticians as far back as the 1960s using physical representations to explain key concepts. One notable example comes from Youden's 1962 *Experimentation and Measurement*, published as a guide for the National Science Teachers Association.[49] In the volume, when introducing collections of measurements, he illustrates the normal law of error with a photo of fifty plants lined up in neat columns (Figure 3.10). You probably know of the concept with its more modern name—the bell curve, or normal distribution. Curiously this is the only photo in the

Figure 3.10 Columns of plants arranged to demonstrate the normal law of error.
Credit: Youden, W. J., Public domain, via Internet Archive.

entire text, and one can't help but wonder who he tasked with collecting and arranging the plants and why he didn't have them create more illustrative photos of data sculptures like this one.

The space of possibilities

The previous ice cream examples start to define a taxonomy of approaches to data sculptures that I find helpful for fleshing out the design space and choices. Academics and designers have offered a variety of approaches to thinking about data sculptures that can be useful for creators to consider in community contexts.

The 2023 book *Making with Data* offers a rich first compendium of data physicalization today from various fields of practice—designers, artists, scientists.[50] The authors' categories are helpful for clustering a breadth of examples, focusing on what techniques were used to create the pieces:

- *handcraft* encompasses hand-made physical data;
- *participation* showcases where people provide or interpret data physically;
- *digital production* gets into digital fabrication technologies being used;
- *actuation* includes projects that move or change their physical shape based on data;
- *environment* highlights pieces that are setting-specific installations built to interact with their physical context.

A research team of designers and computer scientists reviewed almost fifty academic papers about data sculptures in 2022 to generate another breakdown of the design space, describing the contexts on a matrix with two axes.[51] The first indicated whether the data sculpture was specifically designed and connected to the space it was shown (situated) or not (non-situated). The second categorized each as moveable and positionable (unbounded) or fixed and static (bounded).

In one of the earliest academic papers on the concept, by Zhao and Moere in 2008, proposed yet another approach.[52] They begin by defining a data sculpture as a

data-based physical artifact, possessing both artistic and functional qualities, that aims to augment a nearby audience's understanding of data insights and any socially relevant issues that underlie it.

DATA SCULPTURES AS MIRRORS   117

Informed by their context as computation experts, in that paper they offer a design space that varies on one axis from virtual to physical, and on the other from functional to artistic. Extending into questions of the material of representation itself (like the cardboard BYP100 used in Chicago), they define a second space based on metaphorical distance from both reality and the data.

A co-temporal paper from Moere & Patel (the same Moere as in the previous paper) documents a classroom case study broken down with the same taxonomy of symbolism also used in the prior paper—symbolic, iconic, or indexical.[53] These three are borrowed from Chandler's now classic book *Semiotics: The Basics.*[54] Pieces using symbolic representations center abstract translations from the data to the material and form. The BYP100 cardboard sculpture uses a symbolic representation—the bar. The meaning is encoded via our shared understanding of the abstract representation. Pieces using iconic representations utilize some kind of form connected to the data itself. An iconic data sculpture of the budget might have used dollar signs or bags of money. The meaning is conveyed via some metaphor or icon that is related to the data at hand, but not the data encoded itself. Pieces using indexical representations imitate the subject of the data itself to show it directly in recognizable form. An indexical version of the BYP100 data sculpture might have used stacks of actual dollar bills, directly representing the budget amounts. The meaning is denoted by direct representation of the data encoded. These three levels of representation—symbolic, iconic, and indexical—offer another framework for understanding the space of data sculptures being created right now.

Another 2022 review of data physicalization pieces in published academic work offered a glimpse into intent and design methods that are emerging.[55] Common motivations of pieces in that survey included showing personal data, creating artwork, and informing the public. A top desired impact of the projects in that collection was using the physical aspects of nature to improve understanding. The majority used abstract design elements, like Moere & Patel's symbolic representations, and were small enough to be picked up. The survey found changing the size, shape, and color were by far the most common ways data impacted the physical properties.

These various approaches to defining a design space for data sculptures come from visualization designers and researchers, each trying to understand the way data sculptures work. It is worth mentioning that these frameworks aren't generated by artists, leaving us with a gap in understanding how creative practitioners would break down the space within which they work. That said, the mappings can help community groups creating data sculptures make intentional decisions about the media and forms that will be more impactful for them. Is the piece directly connected to the place where it is being shown? Is the data representation literal or abstract? How does the media you use to make it reinforce the data or message? These are the types of questions that can steer nascent data sculptures ideas into effective mirrors for a community, fostering engagement and empowerment through data depiction.

Real world data sculptures

The last few years have seen a wave of data sculptures created to support communication and memorialization related to COVID-19. There was such an explosion of data sculptures during that period that I've taken to calling 2020 the year of the data sculpture. People engage with physical objects differently than flat visuals; recent research is beginning to give us understanding about how. Research comparing virtual 3D charts with 3D printed versions of the same has found the act of touching the chart helped viewers answer questions faster and more accurately.[56] Other work has found that while 3D bars are compared as accurately as 2D bars, spheres are perceived based on surface area, not radius.[57] This type of finding is an important consideration for designers that want to accurately convey comparisons of 3D objects in space. Additionally, work observing data sculptures in social settings has demonstrated how the object itself can mediate understanding complex topics, such as climate change.[58] Government officials, artists, and activists are building data sculptures that embody these research findings, fleshing out the space of possibilities laid out by the frameworks for thinking about them that I have summarized.

3D charts as data sculptures

Symbolic representations, which I call 3D charts, are perhaps the easiest form of data sculpture to imagine someone making, but also in many ways the least compelling and evocative to me. These are simply data sculptures that take a well-known chart type and recreate it in three dimensions. The BYP100 example is one of the few inspirational 3D chart examples I have seen, specifically because it took full advantage of the moveable nature of cardboard. We can look to the very recent history of public health data communication during the pandemic to see a widely viewed example that helps think through what this category looks like.

News media produced and amplified data visualizations daily during the pandemic.[59] "Flatten the curve," a slogan about a line chart, became a mantra. Silly visual data memes circulated across the internet, operating as explanatory analogies, basic puns, and more; data was the became material to joke with.[60] More seriously, as COVID-19 raged across the US in mid-2020, New York state governor Andrew Cuomo emerged as a straight-talking explainer about the current situation. His daily updates were watched by millions, not just those in the state he governed.[61] The *Washington Post* newspaper described how "viewers at home are getting choked up, and people in California and Colorado are texting about him."[62]

In June of 2020 governor Cuomo unveiled a new explanatory device to help shift public behavior, a data sculpture now known as the COVID-19 Model Mountain (Figure 3.11). This was an area chart rendered in three dimensions on an extended briefing room table on one of the largest media stages in the world. The horizontal axis encoded time since the first detected infection in New York; the vertical axis represented the number of positive cases per day. The curve of COVID-19 cases over time is rendered here as a dark green mountain, roughly four feet high and about twelve feet across. The peak, on day forty-two, is labeled with a physical annotation in yellow.

Figure 3.11 Governor Cuomo in front of Mt. COVID-19.
Credit: CC 4.0, NY State Governor's Office.

120 BUILD MIRRORS, NOT WINDOWS

Cuomo stood next to this data sculpture on national TV to try and explain how we needed to take action to flatten the curve.[63] The rationale boiled down to his key statement, that "we don't want to climb this mountain again." Cuomo was using the model to argue for federal support to help New York cope with a potential second wave of infection. In his television briefing he repeated the driving metaphor over and over—"we don't need to climb another mountain," "one mountain is enough," "we don't want to climb a mountain range." The physical data was his central explanatory metaphor. I think this is the most viewed data sculpture ever, probably seen by tens of millions of TV viewers across the United States.

Did it work? I have no formal evaluation of viewers' perceptions of the model at the time, but it certainly was Cuomo's central explanation for days, and I saw images of it echoed across media after that first briefing he held with it. While most viewers did only see the recorded representation of it, they saw Cuomo interactive with it in 3D, walking around it and gesturing at it while standing, as opposed to simply sitting and narrating as he did for the standard charts he showed. The sculpture was met with derision on social media, widely panned for looking like a science fair project.[64] Despite the lack of evidence on its effectiveness, it is highly noteworthy that one of the top public communicators during the spring US peak of the pandemic used a data sculpture to explain what success would look like.

This approach of 3D charts is the first thing many people think of when they imagine a data sculpture; they try to create some representation of a traditional chart in 3D form. In design and innovation settings there is often the additional piece of some kind of novel fabrication technology—3D printers, laser cutters, computer-aided design (CAD)/computer-aided manufacture (CAM). Yet this type of data sculptures uses very traditional ideas of 2D visual encodings, simply projecting them into a 3D space. Branching out into novel materials and non-abstract symbols offers us more approaches to consider data sculptures as mirrors appropriate for community settings.

Artistic data sculptures

On a human scale, the loss of lives in many communities across the world due to COVID-19 was immense. A popular memorial in response sprung up and spread quickly that summer—the idea of setting up empty chairs to represent the numbers of lives lost (Figure 3.12). From Nashville, to Milwaukee, to Philadelphia, to the White House, these makeshift memorials are physical data

REAL WORLD DATA SCULPTURES   121

Figure 3.12 200,000 empty chairs set up for the October 2020 National COVID-19 Remembrance in Washington, DC.
Credit: Ted Eytan, CC BY-SA 2.0.

sculptures that represent the brothers, sisters, mothers, fathers, and more lost to the pandemic. This idea is simple, understandable, and heart-breaking all while respecting that the data represents actual people. It effectively embraces the loss and attempts to provide some spectacle of the collective grieving process.

Large data sculptures like these help audiences understand magnitude and grab attention.[65] The objects and their hand-crafted feel put viewers in a different state of mind when engaging with the data. This artistic approach effectively showcases how important material and construction processes are to data sculpture work.

More artistic data sculptures typically use a mix of approaches to representation. Artists are not immune nor ignorant to the datafication of our times, engaging it as both subject matter and media for work. Some of their pieces do not adhere to the strict idea of varying some physical property based on a data attribute, approaching the representation from a design point of view more focused on the aesthetic and physical qualities of the piece rather than data accuracy.

Over the last few months of 2021 our home studio slowly filled up with 1,659 pieces of stainless-steel cutlery. At just the right time of the evening, a ray or two of sunshine would come in through the window and create dancing reflections off the metal and onto the ceiling. The cutlery clinked and clanked as we moved boxes around, creating a soundtrack to our sorting of forks, spoons, and knives. Cardboard containers were stacked on various tables, many bulging at

122 BUILD MIRRORS, NOT WINDOWS

the seams from the weight. I kept hoping to find a spork buried amid them all, but sadly never did.

The stacks were the result of months of work to collect and re-use the discarded cutlery. Friends dropped off plastic bags on our front porch, neighbors dug through their basements, random people on the Internet met us on street corners for shady-seeming handoffs. My wife Emily and I were collecting these all for a purpose, to build a data sculpture about the impact of the COVID-19 pandemic on food security.

The US is an affluent country, and my home state of Massachusetts is one of the better-off states within the union from a socio-economic perspective. That said, more than one in five families in the state worry about having enough food to eat, with Black and Latino families disproportionately impacted.[66] This is called *food insecurity*—not having consistent access to enough food, or the right kind of food, to meet basic needs for a healthy life. We've worked with local coalitions for years on the issue and saw the growing need around us in early 2020 when businesses shut down in the early phase of the pandemic and our neighbors lost their jobs. Lines at food pantries became far longer, the food support system was overtaxed and running low, and whole new swathes of people that had never needed support before suddenly were forced to navigate systems that were entirely foreign to them. A federal food assistance program (P-EBT) came to their aid relatively quickly, launched in late Spring 2020 by the US Department of Agriculture. P-EBT offered families a debit card that would refill every month to support their grocery needs. The number of families facing food insecurity fell dramatically, but insecurity remained.

The statistics documenting the scale of the need are stark but hard to truly understand. Applications to the Massachusetts Supplemental Nutrition Assistance Program (SNAP) increased more than five-fold as the pandemic started (Figure 3.13). How can you conceive of the idea of 20,000 households in a single week applying for help? It's almost unfathomable for someone like me, who lives in a town of 80,000 people.

Emily and I saw these impacts all around us, in our children's school community and from our friends in the food security coalitions we were members of. We felt the need to do something to spread the word and activate more people to be aware of the challenge, and to help others access support if they needed it for the first time. Surveying these numbers, we found that in mid-spring of 2020 an average of 1,659 households were applying for SNAP benefits every day in Massachusetts. That number, unlike some of the others, is almost

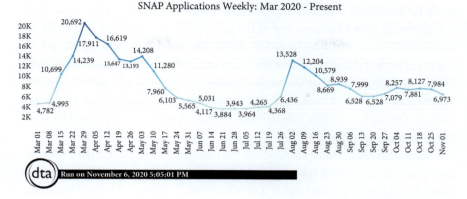

Figure 3.13 Data from the Massachusetts Department of Transitional Assistance on SNAP enrollments in mid-2020.

Credit: Public domain, Massachusetts Department of Transitional Assistance.

comprehensible at a human level, but it still isn't especially humanized. The number is cold and stark and doesn't evoke the level of hunger it represents.

To bring some sense of the true scale of the need Emily and I decided to make a data sculpture—a table constructed out of 1,659 welded spoons, forks, and knives (Figure 3.14). That's how we ended up with a garage filled with boxes of cutlery. Over the next few weeks a table began to take shape, a skeleton of a surface with a leg at each of the four corners. Emily is the welder in the family and toiled away spot welding each piece of cutlery onto the design she envisioned. Occasionally I'd weigh in on design elements, and I certainly welded on a few pieces (likely the ones that fell off first). In the end we had a table about five feet by three feet, standing thirty inches high and weighing around 150 pounds. The piece, which we titled *1,659*, toured the region for more than two years, hosted at urban gardens, farmers markets, art galleries, universities, government buildings, schools, and public events.

Most readers from the fields of data journalism or visualization design would immediately grasp the importance of the number itself, but having it represented in sculptural form as a table opened a new door to understanding the issue and potential responses. One of our first installation sites was a weekly farmers' market in front of the city hall in Revere, a town not too far outside Boston. As we struggled to carry the weighty table up a grass hill to the site, two public works city staff came and offered their help. One of them asked us about the piece, and as we explained what it showed he asked if he could apply for the SNAP program. He had never heard of it, and in fact his question was perfectly timed because one of the goals of that installation was to drive enrollment; there was a table staffed during the market to help people fill in the paperwork needed to receive the assistance. After helping us he went over to chat with the person at the table, leaving us with an early demonstration of how the piece of art could attract people in need to the program. Someone had connected to the data through the sculpture and was motivated to seek the assistance they needed.

This continued at other sites, where the table acted as a data mirror for one aspect of our pandemic trials. These kinds of conversations seldom happen around a bar chart printed out at a public event. Trust me, I've stood in front of many bar chart posters at public events trying to engage the public around data. The physical form of the table, the use of cutlery, the visual design—these pieces combine to offer another way of connecting to the number 1,659. The piece opens a new door for people to relate to the issue the data represents, activating the display space in a way that provokes people to come over and ask

Figure 3.14 The 1,659 sculpture on display at a local urban farm, where it was used to lead educational programing for youth about food access and food security.

Credit: author.

what it's all about. An accompanying sign briefly describes the piece, with a QR code that leads to a website showcasing food pantries, volunteer opportunities, and video vignettes (produced by student Sofia Perez) from interviews with stakeholders in the system during that time when it was so massively overstressed (such as food pantry staff, food rescue volunteers, residents in line to receive food, and others). The table is a data mirror, showing the community more about a problem that doesn't often see the light of day.

Data sculptures as activism

London winters in the late 1700s weren't particularly comfortable compared to modern standards. They were cold, damp, and gray. We can imagine Thomas Clarkson huddled around the fireplace awaiting his guest, trying to stay dry and warm. Clarkson was one of the most prominent leaders of the campaign to abolish slavery across the British Empire. His Society for the Abolition of the Slave Trade began its life after he entered an essay competition on the topic in his youth; moved to action by reading first-hand accounts written by the enslaved. That night in 1788 Clarkson was waiting for his ally, Member of Parliament William Wilberforce. Encouraged by friends, Wilberforce had entered parliament as an independent and quickly emerged as a powerful and eloquent speaker.[67] For the prior decade the men had worked together to try and pass a bill abolishing the slave trade through the UK parliament, failing repeatedly. That night, waiting by the fire for warmth, Clarkson was ready to unveil their newest tool to Wilberforce—a data sculpture.

The two men had formed a bond over their belief that they had a god-granted duty to abolish the slave trade. The abolitionist movement in England produced a variety of impactful public campaigns to convince the population and leaders to outlaw the horrific practices of slavery in the late 1700s. One particularly notable campaign related to how they showcased the manner in which slave carrying ships not only packed humans in unimaginable conditions, but also broke the existing laws at the time that pertained to the amount of people they could transport. This was captured most tellingly in the famous diagram of the Brooks slave ship, which was published in newspapers and plastered across the country to show the inhumanity of the trade. Clarkson described how the "print seemed to make an instantaneous impression of horror upon all who saw it, and was therefore instrumental, in consequence of the wide circulation given it, in serving the cause of the injured Africans."[68]

Within the field of data visualization, the diagram depicting the storage of enslaved people on the Brooks is widely cited as the first civic infographic. It showed how, despite laws limiting the "cargo" allowance to 454 enslaved Africans, the Brooks had in fact been regularly transporting over 600 of the enslaved. It included precise renderings of the ship, with multiple layers and multiple angles. It leveraged that power of data to connote authority and truth. I want to push further than this often-cited example, to this data-sculpture centered part of the Brooks story shared far less often.

When Wilberforce arrived at that imagined cold fireplace meeting in early 1788, forgoing any perfunctory conversation, Clarkson might have reached into his pocket and pulled out their newest tool to fight for the abolition of the slave trade. In his hand he held a palm-sized physical model of the Brooks, complete with depictions of the enslaved on board just like the graphic print (Figure 3.15). He could remove each deck to see images of people painted on the board below. Clarkson had commissioned physical models of the ship itself and was ready to give them to members of parliament.[69] He used the physical model to depict in small scale both the morally outrageous and law-breaking practices of the slave traders. Clarkson must have believed the object itself allowed for inspection and understanding in a way the flat infographic pamphlet did not.

Figure 3.15 A physical model of the Brooks slave ship, created in 1788 as part of the abolitionist movement.

Credit: Wilberforce House, Hull City Museums and Art Galleries, UK© Wilberforce House Museum/Bridgeman Images.

This sculpture was their new way to explain the brutality of the trade. It let viewers hold the data in their hands, to touch the slave transport data regularly collected by authorities. It also allowed the data to touch them, boring into their brains and hearts with an appeal distinctly unlike any other. Wilberforce went on to use the model in the House of Commons, introducing yearly attempts to abolish the trade. They finally succeeded in the 1807 passage of the Slave Trade Act, which banned the slave trade across the British empire.

When we think of communicating with data, often the first images that come to mind are spreadsheets and charts, or infographics and animations, not small wooden models you can hold in your hand. To make a broader impact we have to dispel ourselves of those limits to our creativity and imagination. Data is touchable.

How we interact with data sculptures

Why the recent surge in data sculptures? I attribute it to a broader realization that we need popular data in community settings. Specifically, we need a larger set of techniques to communicate effectively with broader sets of audiences in appropriate and impactful ways. Educator Seymour Papert's concept of *constructionism comes* into play here, specifically how it illuminates the power of externalizing concepts in physically co-created objects to support learning and exploration.[70] This offers us a framework for understanding another motivation for creating data sculptures; that any co-building process is in fact a dialog of a group undertaking to understand the data, construct a story they want to tell, and map that onto a physical representation. In collaborative settings, the process of making a data sculpture is as critical as the end result itself, because the objects embody iterations of understanding and collective sense making. The pedagogy of constructionism can help build critical data literacy.[71]

A number of projects are using data sculptures as participatory data capture.[72] The simplest version is a poll where people are invited to place objects by their answer, building a physical representation of group preferences over time as people "vote." This is often employed at science museums to introduce the idea of counting and related comparative analysis of quantities, or tip jars with two choices at coffee shops. More complicated examples might ask users to work with physical objects, often called tokens, to build some kind representation of their perspective to add to a collective illustration.[73] In public settings visitors can be asked to define themselves by connecting points with thread to create a network diagram or a parallel coordinate visual. The influential Barcelona-based Domestic Data Streamers

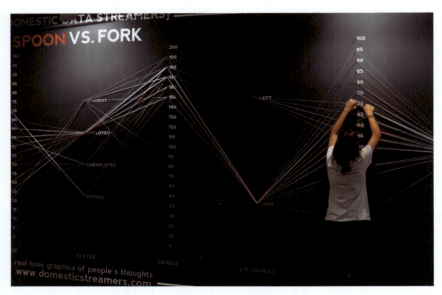

Figure 3.16 The Domestic Data Streamers asked people to answer demographic questions by stringing thread between pins to create a parallel coordinate visualization, such as this 2014 version of their "data strings" installation.

Credit: Domestic Data Streamers.

created many early evocative examples of pieces like this (Figure 3.16), inspiring work like the mural from the Justice Centres Working Group in Kenya. I put my *Just Data Cube*, created for the 2017 Data for Black Lives event, in this category of pieces.

Data sculptures can still be pushed further than the practices emerging today. We can more often build out of materials that match the message, using media that matters and taking advantage of its properties. The materials we choose to create data sculptures with are more than just flat things to be shaped; they carry properties that can be taken advantage of to tell a stronger story when building indexical representations. Additionally, many data sculptures I see focus on the five-second wow-factor, they have very few layers of understanding to dig into beyond the first reading. There is power in the simplicity of the examples here—people just understand them very quickly. I won't argue that building a clear and understandable data sculpture is inferior to building an overly clever one. Still, I think there is more opportunity here. A cardboard chart surprises and delights, but when they come over and talk to you about it, how can you build on that descriptive technique with the further data you want to show them? Could a cardboard box be opened to reveal more data? What physical and material prompts can be put into place to pull people into deeper layers of reading and more interactions with the data?

Use new media for data storytelling

This idea of creating physical representations of data, data sculptures, aligns with many of the central guiding principles of popular data I've already described. Data sculptures lend themselves to participatory co-creation, building on long-standing artistic practices from the arts. Those invitations to participate are welcoming to non-data people due to their physical nature, which is less intimidating than spreadsheet or tool-centered introductions to data.[74] The sculptures provide an alternate path to engaging with data, engaging experiential and practical ways of knowing over propositional. Leveraging public space, data sculptures engage with the power of spectacle to bring data to more popular settings and more popular physical forms. The examples I've highlighted in this chapter offer various models of how to apply these principles via the practice of building data sculptures. The popularity of data sculptures is growing; the Open Data Institute has a catalog of over a hundred pieces of physical data art,[75] and the Data Physicalization Wiki includes more than 300 historical and contemporary pieces.[76]

Consider one creative application of this approach from the field of data journalism. The *New York Times* dug into major donors funding the 2016 presidential election campaign and found that just 158 families were the sources for more than half of the early campaign donations.[77] They took a creative approach to telling this data story in their online edition, using a visual depiction of a virtual data sculpture composed of toy houses and hotels.

Most Americans grew up playing Monopoly, it is one of the most popular board game of all time.[78] In the game players purchase property and charge other players for landing on sites they own, collecting larger amounts of rent from others if they purchase little green houses or large red hotels to put on the sites. The authors decided to use these pieces as a visual metaphor, beginning the piece with a 3D-generated visual of 120 million Monopoly houses stacked in a pile in front of the White House. As the reader scrolls down, they are zoomed into the top of the pile, where a small pile of 158 red hotels sit atop (one for each of the families donating the most). As the article continues, the physical metaphor is used over and over to compare quantities as piles of hotels. Eventually these bridge to satellite imagery of the houses and eventually other photos from the air of where they live (concentrated in a handful of neighborhoods).

130 BUILD MIRRORS, NOT WINDOWS

This visual data story uses the evocative power of well-known physical objects, Monopoly houses and hotels, to create an engaging explanation of a complex topic. While this example doesn't benefit from the properties of form and material in 3D space, it does demonstrate an explanatory physical metaphor based on a sculptural form. The visual designers created a virtual data sculpture and used it as a narrative device to show the sheer outsized impact of the small number of families distorting democratic ideals of power and speech. Creative thinking like this example inspires new ideas for pushing against constraints you might face in your community data projects. The *New York Times* decided to stretch what they could do with on-screen reporting.

Whether you look in your toolbox for low-tech or high-tech approaches, thinking about data as a mirror rather than a window offers very different opportunities and outcomes for production and display. The idea of data as a mirror challenges modern data norms. The idea of a mirror puts opportunity for sensemaking and action in the hands of the subjects of the data, resetting harmful disparities in ways that parallel the goals of the pro-social sector where data is now commonly used. Producing data to capture on tablets, whether clay or computational, for sending off to experts for analysis and decision making outside of a community is harmful and disempowering.

Data sculptures contribute to our growing toolbox of creative techniques for using data in community settings. Research into psychology and perception suggest that data sculptures can invite viewers into discussion more effectively. It is a technique well-suited to public contexts where a complex system is being explained. Museums, schools, newsrooms, libraries, and related spaces can meaningfully employ these practices to further their pro-social, pro-community, and capacity-building goals only if they utilize recognizable forms to represent data. Data sculptures and data murals present two examples of how to do this, learning from methods and media commonly used in pro-social settings and putting the principles of popular data into practice. Each offers practical lessons on the application of the theories in contexts supporting the social good. Taken together, they show how popular data contributes to a growing set of liberatory data practices via using non-standard media to represent data stories. These approaches engage a broader set of media to represent data, and a more intentionally chosen set of methods for engaging audiences around data as a mirror. *Popular data pushes us to break the windows and build mirrors instead.*

4

Create Layers of Reading

Data stories built around charts and graphs can get complex and nuanced quickly, turning off many audiences. You can create layers of reading that help audiences go from simple to complex stories, starting with simple explanations that lead them into detailed narratives. New embodied approaches like data theatre present useful alternatives to reveal complexity and engage diverse audiences in community settings around data without overwhelming them.

A data fashion show in Tanzania

Purple lights fill the spacious hall. Chairs line each side of the large runway that spills out from a curtained stage. A large crowd mills around the room, chatting and glancing at the stage expectantly. TV reporters and news journalists prepare their cameras and notebooks. Well-dressed spectators take photos in front of a backdrop full of sponsor logos. The energy is building up in anticipation of the event starting. The venue has all the trappings of a Paris Fashion Week show, but the location is southern Dar es Salaam, Tanzania. At the center of the stage reads a sign that reveals just how different this is from Paris—welcome to the *Data Khanga Fashion Show* (Figure 4.1).

A runway might be the last place you'd expect to see data on display, but for the same Tanzanian team that produced some of the data murals shared before, this was precisely the surprise and exposure they needed in order to showcase their data literacy and innovation efforts in the Temeke district of the city. The goal? To help youth engage in important data-driven conversation about issues impacting them to reduce gender violence and human immunodeficiency virus (HIV) transmission.

Fashion, culture, and advocacy are connected in cultural traditions all over the world. What we wear operates as a symbol of both exclusion and inclusion in dominant cultures and communities. We protest by breaking the norms of legitimized public expression via clothing; what we wear serves as a sign of

Community Data. Rahul Bhargava, Oxford University Press. © Rahul Bhargava (2024).
DOI: 10.1093/oso/9780198911630.003.0005

Figure 4.1 A model walking down the runway at the Data Khanga fashion show.
Credit: Tanzani Bora Initiative.

our identity and affiliation.[1] Hip hop has used fashion as a vehicle for protest since its early days. The green bandana is a modern symbol of support for reproductive rights and freedoms across Latin America.[2] Palestinian women have been weaving messages of protest and freedom into dresses for more than eighty years.[3]

In the region of Eastern Africa where Dar es Salaam is located the tradition of fashion and protest come together on the khanga[4]—a patterned rectangular cloth with short text written on it. This message, or *msemo*, is often a meaningful saying or proverb.[5] Their use can be traced back to the 1800s, originating to mimic the elaborate clothes of colonizers and to make subtle jabs at neighbors with short quips embroidered on them.[6] In Swahili communities of the region the khanga created an opportunity for women to work their way around social norms that prevented them from speaking about certain topics or in certain settings. In the 1960s khangas were used to protest in secret against British colonial rule, allowing oppressed Tanzanians to make statements in their own language on the clothes without fear of arrest.

When Tanzania Bora Initiative (TBI)—a youth-focused civil society organization (CSO) based in Dar es Salaam—wanted to help youth use data to discuss harmful patterns of gender violence and sexual assault they turned to this rich cultural history; the tradition of designing khangas to make social statements. The program introduced young fashion designers to public data on important social issues and held a competition to design the best *data khangas*—khangas whose designs and text held data-informed messages and iconography (Figure 4.2).

TBI led seventy-five participants through an introduction to data literacy and invited them to design khangas about an issue from the civic datasets provided related to gender equity and gender violence. The primary dataset was the Tanzania Demographic Health Survey, a regularly occurring government-run sampled survey instrument with questions about a wide array of health-related issues.

Back in that purple-lit hall hosting the runway show in 2018, the music swelled and models wearing the winning designs strutted out. The logos in the backdrop of the attendee selfies included PEPFAR (the President's Emergency

Figure 4.2 The winning data khanga design showcases the shocking number of married women who have reported abuse from their husbands (50%). The *jina* (text) states "if you keep quiet, they will make you cry."

Credit: Tanzani Bora Initiative.

134 CREATE LAYERS OF READING

Plan for AIDS Relief), TBI, Data Zetu[7] ("our data" in Swahili), (FASDO) Faru Arts and Sports Development Organization. The audience included the fashion critics, press, and the public. A panel of fashion industry experts served as judges. What did they all see?

Designer Shahbaaz Sayeed shared that "working with data ... allowed me to tell more deeper stories."[8] The data khanga Sayeed designed featured a repeated pattern with three shards of broken glass representing that one of three women in Tanzania had experience sexual violence in their lifetime. Under a central figure of a woman the text reads "Don't judge me for what I've gone through," speaking to the stigma associated when trying to speak up and defend oneself.

Designer Winifrida Touwa showed the same statistic with a design that included three unripe oranges in the center, just one of them peeled. The khanga read "I'm not ripe, don't pick me," speaking to the abusers who attack women at young ages. Touwa noted how she could "use data in my designs to send a message to my society."[9]

The winning design, from designer Danford Marco, was covered in pairs of hearts, each with one filled red and the other hollow (Figure 4.1). This represented how half of married women had been abused in some form by their husband. The design mirrors historical uses of the khanga, but instead of speaking in secret due to governmental repression, it speaks against strong social and cultural norms that require silence about the issue of intimate partner violence. The fashion designs created an on-ramp to learn more about the data that described the issue depicted on the khanga. The cloth could be appreciated as a beautiful artifact of design, or the viewer could dig in one more layer and read the message of solidarity it offered: "if you keep quiet, they will make you cry."

This is not your typical data display, nor the typical group one engages when working with data in public settings. Libraries have shown that most youth still think of data in terms of scientific experiments, surveys, or charts and graphs.[10] Here we see young fashion designers promoting conversations about sexual health, gender equity, and safety. They used the language of spectacle and public performances to spread important messages and create a buzz. Their simple and evocative graphics opened up a conversation that was otherwise hard to engage in, echoing a classic role arts play in facilitating dialog. This program challenged what data is, who it is for, and what could be done with it.

Examples like the data khangas show how using a more creative toolbox can create new approaches using data as an applied community practice. Statistical literacy shouldn't be a barrier for participating in community dialog, and making charts and graphs isn't the best way to reach all audiences. The data khanga project used participatory methods and a traditional material to bring people together around data in the service of addressing gender-based violence. The methods were built around opening doors for new communities to work with data, access to open data about important social health issues, a focus on supporting a threatened group, and a public showcases for arts-based data stories. The raw materials selected were locally familiar, historically grounded in protest, connected to the subjects of the data, and culturally resonant. Their work started simple, leveraging culturally appropriate graphic textile design practices to offer a pathway to understanding more complex and nuanced data narratives.

Create layers of reading

Data visualizations are often paradoxically criticized for being too complex and too simple. Audiences unfamiliar with a topic, or a particular visual way of representing data, aren't likely to engage with complex data visuals unless they are predisposed to be willing to spend time learning. On the other hand, simplistic data representations leave too much unanswered, evoking more questions and confusion due to lack of explanation or context. The key here, inspired by interactive data exhibits in the museum setting, is to consider a data representation that serves multiple types of audiences at once. How do you reach someone that has never heard of the topic your data is about? What about people who know a little about it but are likely to disagree with your core narrative? If a domain expert shows up how will you showcase the more nuanced caveats that always exist? These sets of questions motivate the idea to *create layers of reading*. Data representations can serve multiple audiences by building from simple to complex explanations through design approaches that invite the viewer or participant in, step by step, fleshing out a story layer by layer from simple to complex. An approach that allows for layers of reading offers a short and sweet version of the central story, in addition to a long and nourishing one.

Crafting data narratives that provide for multiple layers of reading is challenging but creates opportunities for pulling in readers and viewers of data stories that might otherwise be turned away by complex data representations. You can tell complex data stories, but with communities who might not know anything about the data being represented you have to start with

136 CREATE LAYERS OF READING

something simple and hold their hand along the way. Like the museum exhibit designer, you have to focus on designing for delight, surprise, and capturing the attention of a potentially disinterested audience. Like journalists, you need to have a compelling headline and also flesh it out in a detailed story.

A team of co-design researchers offer one helpful take on what it means to offer readable data; Schoffelen et al.'s 2015 paper on "Visualizing Things" lays out three approaches for data visualization designers making data public.[11] The first piece of guidance focuses on the importance of attracting people to any data depiction as a design goal, while simultaneously allowing any public viewer agency to decide whether to engage or not. This is a tricky balance that requires experience and iteration to achieve. The second tip is to support making contextual sense of the meaning of the data being displayed. Multiple levels of scale can be included as someone is pulled into a data representation to walk them through layers of reading.[12] Their third advice is to enable and encourage reflective interpretation of the data to understand meaning and impact. They point to the "adversarial" idea of showing non-dominant perspectives, and how that can support viewers taking dialogical and critical perspectives, considering the context and complexity of various stakeholder perspectives.

This three-part approach to making data readable in public offers a useful framework for how to conceive of creating layers of reading in your own community data narratives. You can open your data story with a hook to pull people in, flesh it out by putting that hook in context and connecting to their experiences, and close with connecting to details and impact. This is a skeleton for creating more engaging data experiences, especially in settings where the levels of interest and data literacy vary widely among the audience. Don't be scared to include visual complexity. While often a barrier to engaging viewers[13] it can also support longer and deeper engagement;[14] you just have to build toward complexity slowly.

The constrained toolbox of charts and graphs leaves designers with very few options for introducing simple data stories that slowly reveal complexity and nuance. Created for the sciences and statistics, charts and graphs rely on a single language of visual encoding, one that significantly limits what broadly is accepted as doing data visualization. In community settings we need to understand the history of these visual norms and challenge them to engage people in more appropriate and effective ways on complex topics. We need a broader toolbox to craft stories that use layers of reading to engage audiences from simple beginnings to rich and textured endings.

Problematic norms of data visualization

My first forays into data visualization began in the early 2000s, when I worked at a tech startup writing software for tiny wearable computers. I was building massive visualizations to solve problems in how the custom network of devices communicated, printing out wall-sized posters with millions of colored lines that represented data flowing between them. I was doing data science, but like so many programmers from that era I hadn't been formally trained for it. The growth of data visualization and storytelling as a field has been built on migrants from other domains; people who stumbled into the work much as I did. In response we've seen huge numbers of offerings to professionalize the industry via educational certifications, a parallel push to create norms, and popular training sessions led by experts in the field.

In 2008 I found myself at one of those sessions in a Boston hotel conference hall. Around me were data designers, engineers, and programmers; all curious to learn how to create more engaging and effective data visualizations. The audience was well-dressed, mostly male, young, and quite educated. Patagonia jackets embroidered with ivy league and tech startup logos, MacBook Pro laptops covered with stickers and custom decals, rectangular framed glasses— this crowd felt like my stereotype of the San Francisco tech scene. The hall bubbled with a low murmur of anticipatory conversion, awaiting the entrance of the speaker we had all come to see ... Edward Tufte.

On the list of foundational figures of the modern professionalized field of data visualization you'd be hard pressed to find a more influential figure than Tufte. Trained as a political scientist, he began teaching statistics at Stanford with famed statistician John Tukey. Out of this teaching came Tufte's pioneering work on information design, captured in four self-published books. They are widely considered required reading for anyone working in information design and data storytelling. Chart junk, sparklines, small multiples—these are all ideas Tufte added to the lexicon of the field.[15] His examples are canon; his contributions and approaches are frequently hailed. It is hard to overstate Tufte's impact on the generation of data visualization designers who matured in the 2010s, through his books and through a regular schedule of public talks.

That's what drew me to that bustling event hall in Boston, the chance to learn from the living legend himself. The fancy ballroom was a bit intimidating for me; at the time I was more used to a laid-back jeans-and-sneakers kind of tech startup work environment. My nerves were growing, so after checking in I quickly found a seat. The large stage was curtained, with rows and rows of chairs lined up facing it. The upholstered seats were covered in a generically appropriate fabric, a color I'd call "hotel beige." They all pointed the same direction, toward the well-lit stage. Settling in, I let the anticipatory conversation ebb and flow around me as we all waited. Soon after the appointed time the lights went down, and an announcer called out a simple intro (as if any was needed). Tufte himself emerged on stage to thunderous applause. Dressed in khakis and a white patterned button-down shirt, he proceeded to eloquently talk through the key points of his books, working the stage and the room like an expert showman. He offered dramatic pauses, showed off priceless ancient manuscripts, and shared tidbits about his own personal experiences. I was entranced and engaged; this was like a data visualization nerd rock show.

Two hours later, as I left the grand hall, I finally took a mental step back and began digesting the event. I had shown up curious about how to put Tufte's advice on information design into the messy practice of reality in my job but left with few ideas about how to proceed. I walked away inspired by examples, but with nothing to do about it. I was in a room of hundreds of visualization experts and learners but had no support to find common ground and connect with them on shared problems or possibilities. The event had been built around listening to a "sage on the stage," not the community in conversation that I hadn't fully realized I needed. After the initial high of seeing the legend himself in action, I hit a profound low as I thought about what a missed opportunity it had been.

That experience proved to be a formative event in developing my own approach to teaching data literacy and data representation. Tufte's shadow, and that of like-minded data visualization educators and designers, is long. It posits a set of goals and norms for working with and representing data that were designed for the fields where data originated, not where it is used now. Many of Tufte's examples are incredibly visually complex. In these new settings we need to push for more appropriately informed approaches to working with data, designed specifically for the contexts within which they will be seen. While some can dive into a complex information graphic and are motivated to explore dense visual encodings, others are put off by dense data charts that require significant time to fully understand. Visual complexity scares many audiences away. The current default approaches are not serving a diversity of communities well, but to understand how we must first flesh out what the statistics-centered, science-informed, and math-laden modern approach entails, understand the impact that has had on efforts as data science has spread, and discuss the motivation for a broader set of approaches that are more informed by the context of use.

A narrow definition of data visualization

Data journalist, educator, and designer Alberto Cairo labels this set of norms the *"Tuftean Consensus"*—a set of universal design principles that strict followers of Tufte's approaches have adhered to and put into practice for data visualization.[16] This scientific approach to creating visual depictions of data dominates our current practice, built on the idea of numbers and their accurate graphical depiction via two-dimensional (2D) visual encodings as the best and most appropriate way to understand something. Truth, evidence, and accuracy are important, but there are multiple ways to present data that adhere to those traits.

The key point I want to highlight is that the functional approach embedded in our data visualization practice is limiting because it proposes a very narrow set of norms, built on the scientific method, as the only way to show information accurately. This doesn't allow for consideration of depictions of experienced data, produced from an individual or community's point of view. It allows only for the view of data from some imagined externalized perspective, encoded within the allowed for formats. Social scientists, historians, and feminist scholars critique this singular focus and celebration of scientific inquiry across society at the expense of other ways of knowing. Worth highlighting here is *feminist standpoint theory*, which argues that knowledge is socially situated; it is a product of power hierarchies and marginalized groups are well-positioned to critique it.[17] The visual grammar of charts and graphs limits our ability to include context that situates data.[18]

These accepted rules are broadly baked into the social norms around creating data visualizations and the technologies we use to design them. I summarize the core elements of this consensus in four parts.

First, it strives for a high level of data density. The information designer is understood to begin with a flat, two-dimensional canvas for representing data. Each mark on the page can represent data or not. In *The Visual Display of Quantitative Information* Tufte defines this ratio of representation as "data-ink ratio."[19] He further extends this idea by comparing the number of data points represented in a graphic to the physical area of the graphic itself—a ratio he calls "data density." Adherents to his approach strive to maximize data density and the data-ink ratio.

Second, they design with a minimalist Scandinavian aesthetic. The field of information design is culturally informed and situated. When teaching design students, my favorite example of this is the grocery store—browse the "world" or "ethnic" aisle and you'll find a wide variety of colors, typefaces, and design

140 CREATE LAYERS OF READING

norms that are closely associated with different cultures. The grocery store is a microcosm of the culturally informed visual design practices. Despite this richness, if you scan top papers for academic visualization conferences, or review modern award-winning data stories from journalism and design fields, you'll find a single clear dominant aesthetic at play. The consensus approach is modeled around tidy layout, formal language, and crisp text and visual design. I describe this as a Scandinavian aesthetic, reminiscent of the widely known design style of the IKEA furniture company. You'll see few African design sensibilities like the data khangas showcased in the canonical inspirational examples.

Third, they center an approach of scientific objectivity. Newcastle University journalism professor Murray Dick describes this as the *functionalist-idealist discourse*.[20] Visual data depictions typically speak in the rhetorical voice of science, truth, and fact. Here I'm not trying to argue that science, truth, and fact are bad. However, if a community isn't approaching data from a scientific point of view, then your goals are poorly served by creating depictions of data that leverage scientific norms. You can use alternate techniques from a border toolbox of data representation to construct and convey data knowledge to audiences in appropriate ways based on their context.

Fourth, they almost completely ignore qualitative forms of data. I tend to define data, in modern usage, as factual information that has been produced and stored. One can distinguish between various types, but often data is described as either *quantitative*—counted or measured—or *qualitative*—descriptive. Tufte's work specifically focuses on quantitative information; the word is in the title of his first book. More broadly, the word data seems to be used in the field of visualization as a placeholder for quantitative information. This significantly limits the kind of information we can use, especially in community settings where many react negatively to working with numerical concepts and structures. Engaging a broader definition of data helps us understand why journalists take to data science. They are trained to work with qualitative data extensively, producing it via interviews, validating it via fact-checking, and curating it into written stories. Years of working with data journalists and training them has shown me that these skill sets honed on qualitative data transfer readily to working with quantitative datasets.

In that Boston lecture I attended years ago Tufte described these four core elements in depth, with examples from both modern and historical practice. In 2020 I had the pleasure of working with Elizabeth Borneman, a master's degree student in the Massachusetts Institute of Technology (MIT) Comparative Media Studies program. Her research focused on what role data visualizations do, and could, play in "perspective shifts and communal cohesion."[21] As part of this work, I encouraged her to interrogate Tufte's examples to see whose work was being represented in these canonical industry texts. After reviewing *The Visual Display of Quantitative Information*, she found that his early examples were overwhelmingly created by white men. Of the three women mentioned in the book, two were included as negative examples. Just 3% of examples came from men of color, and none were from women of color. This type of myopic depiction of accepted examples has severely limited our ability to imagine alternative approaches, because we aren't looking broadly enough for examples. Beyond the authors' demographic similarity, almost all of them used paper or screen as the media of production, with 2D images representing data produced by historical hand or modern-day machines.

Truly inclusive data practices require understanding that today's norms are some of many legitimate ways to work with and represent data. Cairo pushes us to reconsider these norms we work with when he discusses the Tuftean Consensus, learning from our peers and their practices in real-world settings to allow for a multitude of pluralistic "dialects" to emerge within the field of data visualization. For these dialects to engage people, we need to consider the history of the dominant complexity in our visual data depiction language and then break from it.

Tyranny of the 2D

As the data khangas and other examples show us, graphs are not the only way data could be represented visually. The small toolbox of standard charts and graphs[22] has contributed to a practice of data visualization that excludes alternate approaches and uncritically embeds a strong set of pedagogical defaults. The prevailing methods and media were not created for the pro-social settings where data is now commonly used and are poorly suited to many of those contexts due to their intentional visual complexity.

The practice of making charts and graphs has created a small library of acceptable depictions of data, a phenomenon I aggressively label the tyranny of the 2D. This relies on the idea of *visual encodings*, an approach to connecting changes in the magnitude of some quantitative data to changes in the visual presentation of symbolic elements. The most often referred to formalization of this idea is cartographer Bertin's 1967 *Semiology of Graphics*.[23] He offered position, size, value, texture, color, orientation, and shape as visual channels that can be varied based on data (Figure 4.3). Bertin argues that we can use

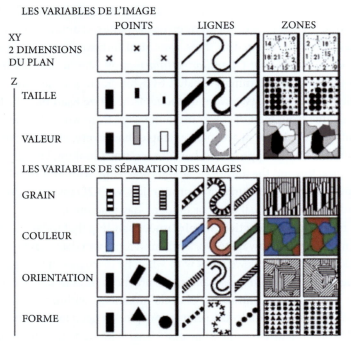

Figure 4.3 Bertin's visual variables.
Credit: Esri and its data contributors. All rights reserved.

marks—points, lines, shapes—to encode information via positional, temporal, and retinal attributes. Most visualization software tools created in the last two decades embrace this approach. If you are familiar with the Tableau Desktop visual analysis tool, you'll recognize this idea from their user interface panels. If you have used lower-level software libraries such as Vega, you'll see how deeply embedded this idea is in their visualization grammar. While Excel doesn't surface the idea, it is behind every setting you can fiddle with on a chart. Two-dimensional visual encodings govern most visualization tools, providing a helpful, but also constraining, toolbox and conceptualization of data representation.

What's the best way to use visual encodings? Many learners look to chart choosers for suggestions on how to get started based on recommended encodings (Figure 4.4). These guides embed the idea that a particular type of chart is best suited for displaying a particular type of data. Decision trees such as these are being built into visualization tools now, via new features that automate chart creation based on algorithmic assessments of characteristics of data you are working with.

PROBLEMATIC NORMS OF DATA VISUALIZATION   143

Figure 4.4 Financial Times Visual Vocabulary Guide educational poster.
Credit: Financial Times Interactive News.

Chart choosers, while presenting themselves as expert guidance, in fact limit the range of possibilities radically. While offering the idea of quick solutions, they don't consider the audience, setting, and goals that should inform the choice of how to represent data. Charts exist within social contexts, yet the chart chooser lives outside any context that a critical data literacy educator, or a practitioner of data feminism, would introduce and interrogate.

The most popular software tools don't offer support for thinking outside of this tyranny of 2D. When you're first starting to work with a dataset on some problem, it is critical to frame an appropriate set of questions aligned with your goals. Data analysis and charting tools like Excel help you dive into a dataset with low friction, but don't scaffold you in identifying appropriate questions to ask. In pro-social settings you need to take a step back and think about what you're trying to achieve. What is the goal of your work with data? What is the social context of the intended audience?

Humans think and understand the world through stories. Background context, strong characters, a clear flow from start to end, steps that lead one into narrative complexity—these are the key elements of the story that help engage us all. Telling stories and making arguments with data is no different. Data exploration and visualization tools like Tableau can be very powerful for

144 CREATE LAYERS OF READING

aiding us in visually crafting data stories, but they don't support the process of coming up with the narrative arc. Which characters appeal to the audience we have in mind? What resolution of our data story will drive action toward the goals we want?

Once you've got a strong narrative based on your data, how do you decide what format to tell it in? This decision has to be driven by a strong understanding of your audience and goals. Data analysis tools like RStudio can't make those connections for you; you have to consider the constraints yourself. Will readers have the visual literacy and geographic awareness that it takes to read a map visualization? Are they predisposed to agree or disagree with you? Are you presenting in a formal setting to an engaged audience, or on the street at a festival?

We've adopted, and centered, a set of data representation technologies and methods that don't address any of the contextual applications of data visualization. Some are working to push the boundaries of this visual language. Journalist and data artist Mona Chalabi creates hand-drawn data visualizations in the playful spirit of explanatory graphic designer Nigel Holmes. Even when celebrated, they are exceptions to the dominant trends. The question of what visual data representation to choose involves so much more than simply what type of chart and which visual encodings to use.

Do charts work?

Why is this small library of charts and graphs so common? Do they work better than other potential depictions of data? Are they universally understandable? The evidence base assessing the effectiveness of this library of common graph types focuses on specific definitions of "effective," and yet is still mixed.

Most evaluations of the effectiveness of data visualization come from the field of computer science. If you argue that visual representation of data is a science, then of course it follows that you should assess it objectively for perceptual accuracies and fallacies. Computer scientists have approached this by studying which encodings are the best channels for accurate representation of statistical quantities for readers. In the mid-1980s, Stanford computer scientist Jock Mackinlay took Bertin's enumeration of visual channels and assessed them for accuracy of perceptual recognition (Figure 4.5).[24] He wanted to understand questions such as how good we are at comparing quantities encoded onto size? Does position transmit differences in data more effectively for readers than density? Results from studies like Mackinlay's created

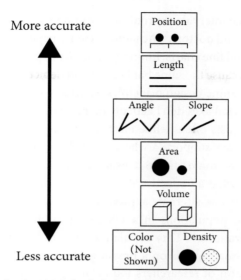

Figure 4.5 Mackinlay in 1986, depicting results from Cleveland and McGill's 1984 work.
Credit: ACM Transactions on Graphics.

guidance that is now baked into many of our tools, which focus on position and length as the primary data encodings. Multiple studies since have found that bar charts are the most used type of these standard charts, in both academic[25] and popular press[26] settings. It turns out that compared to some other forms, the bar chart's encoding of quantity onto length is in fact more accurately perceived by viewers.

Dig deeper and the evidence base becomes murkier. Regarding the Scandinavian uncluttered aesthetic, we see that this format of data visualizations do convey a sense of professionalism.[27] However, informal representations of data increase questioning and engagement in audiences.[28] So you could say it depends. Is your goal to convince audiences that you are correct based on professionalism, or are you trying to engage them with questions and participation? Contrasting with Tufte's claims about "chart junk" hindering the clarity and success of data visualizations, researchers found that charts with "junk" are remembered better than those without and are just as easy to interpret.[29] We do have some evidence that charts with extra labels annotating the story they are trying to tell are more easily understood and hold up more robustly in the recall of the key points over time.[30] However, these same visual encodings can be deployed to deceive readers very effectively in critical settings, such as courtroom juries in the US.[31] Standard solutions like error bars

146 CREATE LAYERS OF READING

to show confidence intervals are routinely misunderstood by trained experts such as professors and doctors.[32] Academic science papers often depict continuous data in bar and line charts, limiting the ability of peer review processes to assess accuracy because the same bar chart could depict many different distributions of data.[33] Some design principles we widely introduce lack evidence, for example truncated axes don't actually manipulate most chart readers.[34]

Computer scientists often define effectiveness as understandability, accuracy, and inquiry task support; a whole sub-field of academic study assesses this.[35] Yet there are some standout examples that take a more contextual understanding of what an effective data visualization is. Computer scientist Evan Peck presents one alternate approach in an award-winning 2019 study, interviewing rural populations in a US city to understand how they perceive data visualizations.[36] This population of non-data experts was shown one of ten data visualizations on a pressing issue: the impact of drug abuse in the US. The team of researchers found that people used a personal lens to interpret the depictions, no matter the visual design. A chart about alcohol provoked a discussion driven by a participant sharing that "I was a functioning alcoholic." A geographic map of drug use immediately drew another viewer to share that "this one is interesting because I used to live in West Virginia." Peck's study, titled "Data Is Personal," provides insights into how the communicative power of data visualizations is closely tied to educational level, political perspective, and personal experiences. The context and lived history of viewers changed how they read even traditional charts and graphs. Visual communication isn't an exact science, perceptions are inextricably linked to individual perceptions and experiences of viewers.

This question of how an audience approaches data is described as *information receptivity* in a 2023 paper from the University of Calgary's Data Experience Lab.[37] Interviewing a small sample of Canadian citizens around data depictions of energy, they propose four nascent categories of information receptivity.

- Data Cautious—people motivated to find information but unsure of their ability to assess it.
- Data Enthusiastic—people ready for data and confident in their ability to understand it.
- Information Avoidant—people resistant to seeing new information in any form.

- Domain Grounded—people highly familiar with data and driven to analyze it themselves.

Their study offers an initial glimpse into situated audience perceptions of data and openness to engaging with data in multiple forms. The research team emphasizes that people moved between these categories quite adeptly; they were not deeply ingrained fixed mindsets. Understanding how audiences approach data is critical in community settings because it drives the selection of approaches for creating layers of reading that draw them into discussions with real world impacts.

These two studies, on rural perceptions of data visualizations and responses to energy data, offer examples of how the field of computer science is slowly beginning to make audience impacts and context central to how it evaluates the effectiveness of a data visualization. We just don't know that much right now about how successful charts and graphs are because the assessment has assumed a scientific approach typically devoid of any understanding of context. The question of "do charts work?" is perhaps better phrased as "for whom do charts work?."

We need more approaches

The methods and goals of pro-social settings, alternate theories of what data is and the role it plays in society, and the limitations of our current 2D systems of representation all point toward the fact that we need more approaches to working with data and representing it in the new settings of data use.

Libraries, museums, journalism, civil society organizations, and civic democratic processes are all wrestling with the statistics-based methods and media of working with data. Those norms are problematic when considered next to goals of those pro-social spaces—engagement, participation, efficacy, impact, cultural continuity, and so on.

One inspirational approach emerging from libraries and CSOs is the use of zines to engage broad audiences on the impacts of datafication. A *zine* is a self-published non-commercial magazine, typically produced with hand-drawn illustration and distributed on paper in small batches. First starting with science fiction in the 1930s, their use grew in the social upheaval of the 1950s and 1960s via underground presses that were used to build networks of counter-culture groups.[38] In the 1980s the punk music scene widely adopted zines for connecting with fans and spreading controversial political

148   CREATE LAYERS OF READING

speech. Throughout that time, communities of color produced zines to publish and share stories that otherwise were not being told by the mainstream press. Zines historically are a tool of communication for marginalized groups, used to share perspectives that run counter to the dominant narratives of society.[39]

Artificial intelligence (AI), Big Data, open data, surveillance—these are all technically complicated and socially impactful issues that can feel daunting to understand. Building on traditions from the zine community, activist groups have been creating zines to introduce these concepts and motivate action to protect the liberties and freedoms their use threatens. Canadian researchers Alex Ketchum and Nina Morena explored the emerging approaches by surveying about a dozen zines that focused on datafication.[40] These included journalist and artists Mini Onuoha and Mother Cyborg's (aka Diana Nucera) *A People's Guide to AI*.[41] That booklet opens with the manifesto that "the benefits of any technology should be felt by all of us." It includes prompts to write down personal reflections, offers case studies of AI in use in technologies already around us, offers metaphors for understanding the design of technologies underneath those applications, and builds a critical literacy through highlighting potential and already real risks. Another example is *Oh, The Places Your Data Will Go!* by Alexis Takahaski, Sophie Want, and Chrystal Li.[42] Their work uses a Dr. Suess-style narrative to trace a fictional selfie photo moving through the vast data ecosystem. They use the familiar form of a rhyming children's story to engage in a complex explanation of the vast technology behind an everyday occurrence familiar to most readers. Various authors are using the format of zines to meet audiences where they are in explanations of how datafication is impacting their lives.

However, the prior examples are a bit more about data than data stories in and of themselves. An example that paints a more useful contrast to the 2D norms I critique above comes from the Carnegie Library of Pittsburgh's 2017 partnership with PublicSource, a local journalism network that concentrates on data journalism. They created a short summer camp for youth to create civic data zines.[43] Beginning with fact-finding in journalistic traditions, they moved to engage embedded biases via discussion, and then to data visualization and storytelling. The local youth participating in the camp analyzed civic data and created data narratives in zine form (Figure 4.6). The results tackled topics from hyper-local demographics to homicide rates, all driven by youth interests. The obvious data-literacy outcomes were paired with efficacy goals for the participants and community network reinforcement goals for the organizing partners. The choice of zines as the key media

Figure 4.6 Some of the zines created by youth at the Carnegie Library of Pittsburgh during the "Civic Data Zine Lab" led by Tess Wilson.
Credit: Tess Wolson.

met the participants and audiences where they were, offering a new and relevant approach into working with civic data. The workshop was part of their larger Beyond Big Data initiative, which has a stated goal of broadening access and creating new on-ramps to data (much like the programs that I participated in at the Dokk1 library).[44] Libraries are consistently at the forefront of engaging people around civic data use in personally meaningful ways. These are zines told data-focused but don't center the format graphical display of information as their primary mechanism for discussing the data.

150 CREATE LAYERS OF READING

Zines are a product of the people, by the people. They take advantage of recognizable and accessible media forms—paper, cartoons—and speak with authority to groups more familiar with peer-based education (as opposed to expert/notice modes of learning). These examples embody many elements of Freire's approach to popular education, putting educational offerings in service of the development of critical consciousness and justice. The creators are trusted community groups, they produce media in multiple languages, the writing is colloquial rather than formal. They mix in visual decoration and comic-style illustrations of key concepts, and they are widely available for free. This echoes work like Gonick and Smith's "The Cartoon Guide to Statistics"[45] and a Microsoft research project into automating the creation of data-driven cartoons based on local news.[46] Comics and cartoons are legible forms for representing data and statistics to many audiences. Zines offer another low-tech approach to meeting people where they are with data stories.

These zines are an example of data literacy and storytelling that is appropriately designed for the context of use. They contribute to a broader toolbox we can pull from when working with data in settings focused on empowerment, engagement, education, or efficacy. They demonstrate how to push beyond the norms entrenched in our standard approaches to working with data.

Act out your data

In early 2023 I found myself walking back and forth around a Northeastern University theater with ten other people, following the instructions of a trusted facilitator who regularly instructed me to change my speed and direction. He said to silently follow someone, I followed. He said to change our pace from super slow to almost running, we all obliged. I did all this with a slightly bewildered look on my face because this was all new to me. I was never a "theatre kid" growing up, so I felt uncomfortable joining physical theatre warm-up exercises like this one. One thought coalesced in my mind as I continued walking to and fro, trying not to run into the other participants ... what had I gotten myself into?

My journey to that studio began four years earlier when I started to brainstorm the idea of using our bodies to represent data with my MIT Media Lab colleagues Victoria Palacin Siva and Laura Perovich. All working on creative approaches to bringing community together around data, we were motivated by shared questions of how to create engaging, impactful, and fun experiences. Victoria was already developing playful participatory civic data processes as part of her dissertation work, and Laura was in deep collaboration with community partners to build participatory physical data experiences related to environmental justice. Our meandering interests and conversations coalesced around questions of how movement could open new doors to working with data for learners who didn't gravitate toward more technical introductions. Could embodying data engage people that learned in different ways? Would movement-centered productions based on data lead performers to more critical perspectives on datasets and their use? Would performing a data story change how the performer and audience thought about those represented in the data? Could embodying data be an approach to putting data back in the hands of communities to use for their own purposes?

With that many questions there was only one path forward—we had to try it out and see what happened. Through multiple rounds of workshops with artists, community members, and students, we quickly became delighted, confused, and intrigued by what participants created. These led to more formal explorations into the idea of data dance and data theatre. I found myself drawn to the latter, the question of what the combination of data and participatory civic theatre practices might offer to settings focused on engagement and deliberation, two goals of the pro-social spaces poorly served by the tyranny of the 2D.

A year later at Northeastern, I found like minds interested in pursuing this further in colleague Jesse Hinson and undergraduate Amanda Brea. They brought deep theatrical knowledge to my naive understanding, and we created more focused experiments with talented undergraduate theatre majors based on the initial activity designs I had sketched out. Let me illustrate with a retelling of one scene created by activity participants. The small group began with a three-page handout of data about eating norms and preferences in immigrant communities in the Boston area, with the prompt to spend fifteen minutes reviewing and discussing and then designing a short scene for performance back to the larger group. To my surprise, the scene this group devised was a debate between fictional community members, arguing that the dataset about eating patterns was reductive, stereotypical, and offensive. They closed by sharing the moral of their story—"always be critical of the data you are looking at, especially if it is not your own." This group had read through the data, clearly had critiques and questions about it, and decided to design a short theatrical debate that challenged it. I seldom see this kind of critical thinking about data in my workshops with learners new to working with data. This

152 CREATE LAYERS OF READING

example, along with other findings from our research, suggested to me that the theatrical activity was creating a novel space for critical thinking about the data, context, and culture.[47]

Simmering on this second round of experiments in data theatre, a year later I found another excited collaborator in Dr. Dani Snyder-Young, who introduced me to the practice of Research-based Theatre (RbT)[48]. Created to help various types of researchers connect their practice to the public, RbT uses theatrical techniques to bring research to life and connect it to real social contexts. The process is built around collective reciprocal authorship of scenes that depict research findings in contextual and understandable ways via dialog and performance. RbT has shown strength for showcasing diverse and marginalized perspectives, offering an intriguing example of community-engaged theatrical practice related to academic communication. An early example of RbT comes from the 2002 work of cancer researchers Gray and Sinding, who created two theatre pieces built on qualitative data from interviews of cancer survivors for performance in hospitals and public venues.[49] They turned to theatre to try to better communicate the emotional experience of the patients, building on practices of verbatim theatre. They performed the results of their research, opening up doors to new audiences and impacts.

Spurred by examples from RbT, we launched a new effort with other faculty interested in what data theatre might be. This is how I found myself awkwardly walking around a rehearsal studio in 2023, uncomfortably and excitedly out of my element. The others walking with me included Dr. Snyder Young, and brought together a talented team with expertise in civic service design (Dr. Michael Arnold Mages), theatrical devising (Prof. Jonathan Carr), lighting design (Prof. Oliver Wason), participatory public planning (Dr. Moira Zellner), acting and directing (Prof. Victor Talmadge), and RbT (Dr. George Belliveau). We collectively designed the workshop to try out the idea of creating quantitative data theatre with a community partner.

Dr. Zellner brought a pre-existing partnership with LivableStreets Alliance of Boston that proved ideal for this collaboration. LivableStreets works on advocating for more urban green spaces across the city, in concert with the local community and the city planning department. They shared data with us, and we designed a two-day workshop to devise and stage a short theatrical piece. Our interdisciplinary team each contributed an activity to this workshop, from the walking warmups described earlier to sticky-note brainstorms, from data handouts to evocative staging. We broke through disciplinary language barriers, for instance realizing that *tableau* in a theatrical setting refers

Figure 4.7 Students performing a data theatre piece based on data from Livable Streets Boston.
Credit: Author.

to a stage picture made with actors' bodies, not a piece of data analysis software. We explored how the idea of the *viewpoints approach* to theatre making was very akin to the idea of visual variables in data visualization, offering a set of channels that data might be encoded onto by changing properties based on data values.[50] In the end, the group produced a ten-minute piece that we staged internally, performed by a set of talented undergraduate students.[51]

Writing about a theatrical performance is challenging; this text format makes it hard to capture the real experience of being in the room (Figure 4.7). That said, I'll do my best to summarize how it played out. The play opens with a couple walking down a path through an imagined open green space in the city, admiring the scenery while unintentionally slowing down a jogger coming up behind them. At various points different micro-scenes play out in the park.

- Sitting on a bench, a local resident shares "I need the contact with the green (nature) to keep me balanced."

154   CREATE LAYERS OF READING

- A young woman looking for an apartment to rent has an exchange with a real estate broker, reciting local rental unit costs until she collapses from the weight of realizing she can't afford to live near to the city.
- Another passerby, a young man of color, walks through and looks at the white jogger derisively, sharing "So here we are, the long-awaited sequel to white flight; white return."

We began with quantitative and qualitative data from LivableStreets about green space access and attitudes in the Boston area. Expanding from there, the dialog was built from direct quotes in research papers on public perception of green space, research on gentrification, and poetic writing about nature. The core theme was pulled from a tension LivableStreets wrestles with between the push to build green space and the gentrification and displacement that often comes after. Various physical movements and body positions were embodied encodings of data from census and other more local data sources, including data produced by LivableStreets.

Within the team we found quantitative and qualitative data to be fascinating inputs to the theatrical devising process. It was source material that actors and the director could work from without significant rethinking of their normal processes. I found it intriguing to see this group re-humanize the data and be so careful and attentive to not speak someone else's words as their own. The questioning and back and forth around the data was so different from typical data science processes I have led in other workshops and teaching settings. Creating data theatre was empathetic, reflective, and respectful of the material and topic at hand. Unlike many data visualization challenges, the team didn't approach data like some shiny bauble to try some new methods on. The data theatre workshop presented a stark contrast to most of the work I do with data in computational and statistical spaces. We were expanding our toolbox and opening new doors for telling detailed stories that start with simple invitations for audiences.

Data theatre as embodied learning

A history of related work on embodied learning, participatory theatre, and civic engagement underlies our ongoing development of data theatre. Beyond interacting directly with the world, your physical body is a powerful tool for understanding and representing your perceptions and mental models. In the West, our general understanding of the human self has long centered on the question of how the mind and body interconnect. Perhaps the most well-known historical framework that has influenced our thinking is from seventeenth-century philosopher René Descartes. His *mind-body duality* is understood as the idea that the mind and body are distinct and separable entities. Ask anyone today and they would probably describe our full selves as some kind of melding of mental and physical entities, a modern manifestation of this idea of dualism—rationality in the mind vs. emotions in the body. That said, many popular texts over the last few decades have sought to reframe our understanding based on new medical research.[52] To summarize and vastly simplify, it turns out there isn't very much of a duality here; our mental reasoning is in fact driven very directly by body-based stimuli.

Psychologists and educators have been exploring this understanding in more depth to better leverage the ways we sense-make in the world through our bodies. Many formulate these into taxonomies of intelligence and memory, some of which have become popular enough that you might have heard them—visual learners, muscle memory, and so on. More than fifty years ago, arts educator Viktor Lowenfeld described a body-centric way of knowing as *kinesthetic intelligence* to get at how we use our body to express or understand ideas and experiences.[53] It is no surprise that these ideas emerge from early childhood education, a period where we explore the world more actively through all our senses—taste, smell, touch, feel, sound—not just the visual that dominates our work on data representation.[54] Work a decade after Lowenfeld's book found evidence that visual and kinesthetic short-term memory are processed in distinct regions of the brain.[55] More recent scholars have documented how embodied experiences support learning,[56] improve long term memory,[57] and increase empathy.[58]

Think of your typical schooling. It probably centered on abstract, mind-based understanding of material. Yet there is growing evidence for, and engagement with, body-based approaches to learning.[59] Educators have argued that how we learn and know is rooted in the *somatic phenomenon* of the learner—the physical experiences they undergo and create while in learning settings.[60] From the perspective of child psychology, body-based learning builds on constructivist (Piaget) and constructionist (Papert) pedagogies. This is the learning of being hands-on and active in the constructive learning experience via the creation of externalized models (in this case, movement). Using your body to act as the data in this way is not only fun, but connects to Papert's concept of *body syntonicity*.[61] He coined this term to describe how children would predict the movements of a real-world programmable mobile robotic turtle by

156 CREATE LAYERS OF READING

imagining they were the turtle itself.[62] Over and over again he found young learners using their physical bodies to digest, understand, and model their learning in the activities he was running. Their bodies externalized understanding of how each command they gave their robot would work, giving physical form to the planning process so they could work on it collaboratively. Further work from museum studies has revealed that children and adults recall movements they do themselves better than ones they observe, lending more support for the power of physical movement.[63] A team of academic researchers working in public schools has demonstrated ways that arts-based embodied approaches to engaging with data can build student ability to critique data arguments (their *informational inferential reasoning*).[64]

Using bodies to represent and interpret data can create a very simple and relatable way into more complex data narratives. The concept of body syntonicity offers learners a non-visual and concrete (e.g., non-abstract) approach to thinking through data. This avoids issues of disparate access to training and resources for computation-based approaches. The concept of somatic memory offers embodied data storytellers another approach to creating impact; body-based motions live on in different ways than exposure to visual data narratives. The empathy that can be cultivated through participation in, or viewing of, a theatrical performance opens a door for audiences to a data story that engages emotional and rational, physical, and mental responses. These are all potential benefits that embodied approaches to data bring to the table for community settings.

Many of these ideas connect to a concept of *embodied cognition*,[65] building on foundational work from Lakoff and Johnson on linguistics, gesture, and thinking.[66] They argue that mental models of abstract concepts are developed in our sensorimotor cortex, built on understanding of physical movements. As one example, developmental psychologist Susan Goldin-Meadow has found experimental evidence for gestures in learning settings being better predictors of understanding and concept development than spoken language.[67] One of her experiments asked children to explain whether water poured from a tall and thin vessel into a short and stout one is still the same amount, a standard way to assess a classic development stage. Amongst participants that (incorrectly) verbally said the amounts were not the same, many actually used their hands to explain in a way that didn't match; their hands actually indicated the amounts were the same. When introduced to the idea of conservation of volume afterward, those children who had a verbal-gestural mismatch picked it up much more quickly. The movements of their hands were predicting their ability to learn the concept. Related work found that the same phenomenon showed up in learning mathematical addition; children who had mismatches between their spoken answers and their hand movements improved more quickly. The same result has been seen in real classroom settings outside the lab.

The young and the old use their gestures and their body to understand new concepts.[68] We act out our learning and understanding physically. American colonizer Gamaliel Smethurst, while in the captivity of Micmac chief Aikon Aushabuc in the late 1700s, wrote how the chief used his hand to represent a map of his native region while discussing the ongoing geopolitical conflicts, making

> almost a circle with his forefinger and thumb, and pointing at the end of his forefinger, said there was Quebec, the middle joint of his finger was Montreal, the joint next the hand was New York, the joint of the thumb next the hand was Boston, the middle joint of the thumb was Halifax, the interval betwixt his finger and thumb was Pookmoosh [the place they were], so that the Indians would soon be surrounded, which he signified by closing his finger and thumb.[69]

Smethurst describes chief Aushabuc using a hand-based model that supported building shared understanding of the geographic data they were talking about. He began by conveying data points of cities and land seized by colonizers, then pivoted via movement to convey the narrative of being surrounded and the future losses he foresaw.

The growth of data-connected learning standards has led educators to design activities such as bodily data sorting and physical infographics.[70] The core connective thread is providing the opportunity to socially construct an understanding of a phenomenon by embodying it. In this practice the act of bodily participating is the construction of knowledge itself.[71] Statistician Brian Joiner had students creating *living histograms* as far back as 1975, lining them up to represent bars of a histogram with their bodies (Figure 4.8).[72] His primary stated goal was to "help make statistics come more alive to students who are not numerically or algebraically inclined."[73] While simply a 2D histogram chart when viewed from afar, Joiner was taking advantage of the insight that we learn differently when we do it with our full bodies.

More contemporary educators and researchers are connecting the body and data literacy. The International Society of Learning Sciences convened a symposium on *Embodying STEM* in 2021 to explore research using dance to introduce topics in the sciences, technology, engineering, and math (STEM).[74] Their key takeaways included the idea that using dance to introduce STEM

Figure 4.8 Joiner's living histogram of students by height, showing a bi-modal shape due to inclusions of all genders.
Credit: Joiner, B./Creative Commons via JSTOR.

concepts "can broaden access to learning and engagement."[75] University of Colorado School of Education professors Sommer and Polman use bodily data sorting activities like Joiner's.[76] Education researcher Kayla DesPortes and a broader team from New York University led engagement with public schools to explore this in more depth.[77] Through co-design processes with teachers, they created a data and dance unit for middle school students. Students created motions to reflect data values and their interpretations. The research demonstrates how learners build their own embodied understandings to represent graphs and data through dance, revealing new paths into data learning and new understandings of the context and impact of the data.[78] Embodied approaches open doors to engaging with data that appeal to newcomers and learners, giving us a new pluralistic tool to use for bringing diverse groups of people together around data.

Learning from a commercial

While watching television one day a few years ago, an advertisement came on that caught my attention, a surprising example of performative and participatory data embodiment in the commercial setting. A kind-looking older man popped up asking me and all the other viewers a question: how much money did I think I'd need for retirement? This question isn't one that most young people are interested in, but over the next forty-five seconds or so the commercial played out a powerful example of engaging people around data by asking them to represent it with their own bodies and motions.

The host of this advertisement was psychologist Daniel Gilbert, who has spent most of his adult life trying to understand how people make decisions, with a particular interest in how we consider the future implications of our current choices. From 2013 to 2016 he appeared in a series of award-winning commercials with the investment and financial services company Prudential, targeted at encouraging potential customers to overcome psychological barriers to retirement planning.[79] The commercials supported Prudential's *Bring Your Challenges* ad campaign, produced by the Droga5 creative agency. The ad I saw was called the *Ribbons Experiment*.

You should watch it online to fully understand the relevance, but for those of you without a device at hand I'll describe it. It opens in a field in Savannah, Georgia, with a group of people walking up to the camera. The field is painted with concentric rings, each noted with an age. Gilbert narrates how they asked people the key question—"how much money do you think you'll need to retire?." Various participants share their estimates, and each is given a yellow ribbon to show how long that might last (the higher their estimate, the longer the ribbon). What follows is a montage of people anchoring one end of their ribbon in the center of a giant circle on the grass and then walking out past concentric rings on the ground labeled with ages until their ribbon runs out. At that moment, when their ribbon runs out, we see shock and surprise at learning the age at which they are projected to run out of money. From above, the results resemble a radial column chart of yellow bars. Gilbert steps in again to tell us about how hard it is to imagine the future and how much money we'll need. We next see participants given blue ribbons, the primary brand color of Prudential, and they walk out in a similar way, this time making it all the way to age 105 as the commercial ends.

I share this story not to encourage you to pursue retirement planning with Prudential, but to highlight this embodied data presentation.[80] The *Prudential Ribbons Experiment* ad is well produced, shot, and explained, but I want to hone in on one moment of the large segment to talk about why it strikes me as an inspirational example about embodiment. Rewind my narration of the commercial in your head back to the moment when the yellow ribbon runs out. Pause it there. Can you see the frozen image of a woman holding the very end of a ribbon? She was just jerked back a little because that ribbon ran out. This physical sensation, literally feeling her motion stopped by running out of money, offers an experience of the data story no chart ever can. She looks

160 CREATE LAYERS OF READING

down at which ring she has made it to, sees age seventy, and her face fills with worry and confusion. The message has hit home, she is fast-forwarded physically and emotionally to a future where she has no money to continue living. The psychology at work is body based. I've watched this video dozens of times in workshops as I explain the idea of embodying data and every single time it makes me want to check on my own retirement savings and goals. As a video, her experience is transformed into a performance for the viewer; I observe what happened to her and immediately empathize with the emotional experience, creating a way of knowing the data that complements the visual depiction in the radial column chart. I haven't embodied the data myself, but experienced the knowledge she is building via the viewing of the performance.

Embodiment is a powerful tool for telling data stories, one that adds creative new dimensions to our toolbox of popular data in practice. While the context isn't particularly liberatory, the ad pushes harder on the current two-dimensional representations of data we use, and how we value propositional and abstract ways of knowing. Embodying data helps us humanize it, a critical educational goal.[81] My early experiments in creating data theatre utilized this power in concert with traditions of participatory theatre as a radical tool for social change and civic deliberation.

Learning from participatory theatre

To understand the pedagogical foundations critical to data theatre I'll return to the legacy and work of Freire and his approach of popular education. Freire's *Pedagogy of the Oppressed* inspired another Brazilian to imagine how theatre could be put into use to transform society intentionally and directly.[82] This director, playwright, and theorists' name was Augosto Boal, author of the accordingly titled *Theatre of the Oppressed*.[83] It is hard to overstate Boal's impact on participatory theatre practices, all of which emerged within the repression and turmoil his country experienced, including multiple military coups and large groups of artists being exiled.

A first model he developed was the idea of *Newspaper Theatre*, which transformed the regular news into theatrical pieces.[84] The performances changed each day (partially to help avoid censorship) and focused on re-contextualizing reported news to help audiences understand what was left out or distorted by the official state news agencies. Exiled from Brazil by the military dictatorship in the early 1970s, Boal went on to develop *Image Theatre* to overcome language barriers by shaping the human body to represent experienced oppressions. Another approach he created was *Invisible Theatre*—planned performances that took place in public space and appeared to be natural interactions but were in fact acted and improvised to highlight oppressive practices otherwise left overlooked. His theatrical innovations continued with *Forum Theatre*, a spectacle that would invite the audience into the performance by pausing it and breaking through the *4th wall*, speaking directly to the audience and prompting them to reflect on the situations being performed.

Boal's theatrical ideas serve as a still-relevant and still-practiced example of socially constructing knowledge and understanding via embodiment. Earlier work by Bertolt Brecht in Germany was pushing back on the idea that theatre had to be escapist, his "dialectical theatre" built a practice that was breaking the 4th wall and working harder to embrace critical reflection in the audience while viewing a performance. Brechtian theatre focused on meaning making through engagement of potentially opposing perspectives.[85] The approach of *epic theatre* is a common root for many of these practices, built around the idea that theatrical practices can provoke the audience to reflect on the world as it is, rather than suspending disbelief and entering an imagined world via the performance.[86] This is a very different type of theatre than escapist musicals you might see on Broadway, or even performances about socially relevant or controversial issues. This is a theatre built to interact with the public and help them think through ongoing events to support building their ability and desire to make a change.

Boal's approach, like Freire's, was grounded in the real-world problems experienced by their communities. It was built as a popular theatre form intended for those in liberation struggles. They explore oppression in many forms—internal, external, individual, and social. The key goals included entertainment, education, and emancipation. This has birthed several other participatory theatrical approaches. *Playback theatre* is one example, bringing participants to share traumatic experiences which are then "played back" to the broader group. It has been used to promote compassion and understanding in medical education,[87] healing, and reconciliation for post-war communities,[88] mental-health treatment,[89] and recovery for adolescents in refugee camps.[90] Playback theatre aims to create a deeper, multi-dimensional understanding of experiences—building empowerment and engagement in decision-making. Verbatim theatre,[91] and image theatre[92] are two examples that have grown from this common theoretical core and been successfully used for empowerment in communities. Theatre has a rich tradition of using body-based approaches to dig into complex stories, putting presentational and experiential ways of knowing front and center.

162   CREATE LAYERS OF READING

Boal's Theater of the Oppressed approach continues to have worldwide influence. Contemporary creative practices he has inspired include those coming from the top-down and the bottom-up. Antanas Mockus, mayor of Bogotá in the 1990s, was inspired by these methods to use mimes, games, and the arts in order to develop a safer civil society, reduce crime, and prevent traffic deaths.[93] Throughout the same decade, artist Suzanne Lacy staged elaborately produced dialogues between youth and police in Oakland, California, to address the profound lack of trust on both sides.[94] In 2007, Paul Chan worked with the Classical Theater of Harlem to stage the play *Waiting for Godot* in the Ninth Ward, New Orleans, the neighborhood that was devastated by Hurricane Katrina.[95] The production involved art seminars, educational programs, theater workshops, and conversations with the community.

There are a handful of examples at the intersection of data and theatre to draw from. One of the most well-known is German documentary theatre group Rimini Protokoll's *100% City* series.[96] Each performance invites a hundred residents to physically visualize data about the city they are in. The piece, performed since 2008, asks a question about their demographics and prompts the audience/participants to move their bodies to varying places on the stage to indicate their answers (akin to Joiner's living histograms). One of their primary goals is to explore how to make data more personally relevant. Their work engages emotional involvement from the participants and audience.

Another example comes from British artist and activist Ellie Harrison, who in 2019 staged her *Bus Regulation: The Musical*, in which she leads a troupe of performers acting out the history of public transit planning in Manchester. Following up on the successful feedback with versions in Strathclyde and Merseyside, the trilogy uses public transit data to highlight public health issues in each of the post-industrial regions. This all happens on roller skates, lending a delightful comedic air to the otherwise dry topic.

Visualizer Federica Fragapane has created theatrical performances where data about the script and actor movements are projected onto the stage, adding another layer of reading about the play itself.[97] This type of meta-depiction is evocative, but ultimately aimed at different goals than my work (participation, engagement, empowerment, and so on).

Even more relevant to the idea of data theatre is a 2015 piece from artists in residence at the New York City Museum of Modern Art (MoMA), titled *A Sort of Joy*. Data artists Ben Rubin, Jer Thorp, and Mark Hansen reviewed the open collections data released by the MoMA and designed a series of "narrative boardwalks" through the data.[98] They created a performance in the gallery space where a team of six performers read and sang out text manually and algorithmically extracted from their collections data, offering a dynamic and curious experience for visitors surprised to see the space come alive in a new way (like Boal's invisible theatre). The performance was all constructed directly from the database, with a particularly playful incorporation of cursing, singing, artists names, and more. It particularly brought alive issues of gender, with an act that involved two women circling a group of men reading off male first names for minutes until one brief female name shows up and they quickly read it aloud in response. Another few minutes of male names and then the women interject again. Their piece demonstrates concretely another approach to how data can be performed and embodied in theatrical ways, expanding the toolbox for data representation in community settings even further.

Community data theatre

These educational and artistic approaches to inviting people to use their body to sense-make are inspirations for how, and critically why, to think about a concept of data theatre. My initial work resulted in two short prototype activities: Embody a Dataset and Make a (Data) Scene.

The Embody a Dataset activity invites teams to review a small data handout and then choreograph a larger group into some representation of that data. Given data about ice cream consumption in the US over time, one team led a group in performative eating of ice cream. We were instructed to hold up a fake ice cream cone in front of our mouths and chant "nom nom nom nom." As they called out years and quantities, we were instructed to change the speed of our eating and volume of our nom nom-ing, faster and louder as the per capita consumption rose to its peak around 1945, and then steadily decreasing to plateau over the last few decades.

The Make a (Data) Scene activity pushes further to devise a short scene based on qualitative and quantitative data. Small teams are given multi-page data handouts and told to devise a five-minute scene to present back to the group. These can be dialogue-based, act out invented scenarios, tell back-stories, involve movement and props—whatever fits the context.

Initial observations from iterating on these activities led me to wonder more about whether theatre created space to connect with the people represented in the data. Quantitative data visualization work is often critiqued as dehumanizing, especially when working with very serious subjects. Aggregation, and the basic fact of representing real people with numbers, separates the data from the humanity of the lived experience represented by the data. We wondered if acting out the story you see in a dataset increased empathy toward the people the data are about.

To illustrate what we saw, I'll summarize another short scene created by some of our workshop participants for the Make a (Data) Scene activity. In this case, the group was given a multi-page handout of qualitative and quantitative data about food insecurity in the Boston area. The group looked at the data and noticed that transportation was a significant barrier to food access in our area and noted how that impacted certain groups more than others. They offered their interpretation of this data as a scene in three parts. It opened with an individual walking slowly down a sidewalk, occasionally doubling over and clutching in exaggerated ways at their stomach to mimic pains of hunger. The second part had them stop near a bench, frequently checking their watch and a wall poster as they waited for something. In the third part another person, acting and dressed as a bus driver, pulled up in front of the first. Their expression and pain eased, turning to relief and gratitude as they boarded the pantomimed bus and set off toward their destination.

After the performance, our debrief discussed their takeaways and intentions. This group spoke specifically about representing the pain of hunger and the annoyance of waiting at a bus stop—they took the aggregated abstract knowledge of the data and translated it back into experiential knowledge that brought empathy to their interaction with the data. As the audience, we took away an understanding of the experience from its performance. This short example illustrates early signs of how data theatre can rehumanize data that is so often aggregated and separated from the lived experiences of those the data claims to represent.

This historical context, experimentation, and collaboration, shows how data theatre can operate as a new tool for bringing people together around data in novel ways. Specifically in civic spaces, the idea of presenting a dramatic piece that is devised based on data, and then leading facilitated conversations that reflect on it, is very evocative. This proposal stands in stark contrast to the more standard form of government/community meeting, where a PowerPoint presentation is followed by a question-and-answer session. This technique opens new onramps for the community to learn the practice of data storytelling, creates space for empathy with the subjects of data, and engages people in critical thinking about data and its application. The theatrical performance can convey the texture of a complex story through relatable scenes and dialog, rather than requiring a detailed understanding of visual encodings and graphical representations. The performances offer layers of reading that open a door for simple understanding or deep reflection on complex datasets presented in narrative form. This body-centered, arts-based, approach offers an alternate way to bring diverse publics together around data that focuses on outcomes, not the data itself.

Build to complexity

Data stories can be complex, full of nuance and detail that make simple take-away messages hard to craft. The idea of creating layers of reading offers us an approach to building narrative data experiences that hold the hand of the audience, leading them from simplicity to complexity for as long as they are willing to engage with the experience. You can reach various audiences, bridging the core challenge of speaking to diverse communities at the same time.

We experience the world as full entities of body and mind, even if the ways we process signals and phenomena from each might be constantly under construction. Our bodily experiences mediate how we understand the world. Embodying data opens a door to a whole new language of data-driven experience, emotion, and impact. It operates beyond the visual, affording new opportunities for data storytellers to design for emotion, movement, performance, and more. Telling data stories through theatrical performance offers a more appropriate form for pro-social settings, further expanding our toolbox for data storytelling.

5

Open many Doors

The term data "visualization" reveals just how central the sense of sight is to modern norms. Yet people engage with knowledge in a variety of ways, and through all their senses. Diving deeper into our senses of hearing, smelling, and tasting adds more radical approaches to the expanded toolbox for creative data representation in community settings. We can build multi-sensory data experiences using culturally resonant techniques to engage communities that are otherwise left out of formal visual-focused data processes.

A multimedia exhibit in Arizona

The area around the small town of Hayden-Winkelman is altogether foreign to east-coast residents of the US like me. Located in the south-western state of Arizona, about an hour north of Tucson, the town is dotted with remnants of the historical mining industry—a smelter behind the school, large piles of slag full of metal byproducts, a nearby gigantic open pit mine that dwarfs any car passing by. The environmental leftovers of the local resource extraction industries have impacts beyond just these visible geographic scars. Residents grew up around acid mine drainage that made the water in their neighborhood run green; air quality so bad they struggled to easily breathe some days.

Environmental scientist Dr. Mónica Ramírez-Andreotta created Project Harvest to work with communities like Hayden-Winkelman, to "address injustices through the democratization of science."[1] Supported by the National Science Foundation (NSF), the project's goal is to understand the pollutants in collected rainwater via empowering collaborations with residents across Arizona who live near old or active mining sites. Partnering with local groups for community outreach, they recruited a team of over 150 residents living in four mining areas to regularly collect samples of rainwater on their property over the course of multiple sampling seasons. These samples were then tested by the Project Harvest team to look for contaminants.

Community Data. Rahul Bhargava, Oxford University Press. © Rahul Bhargava (2024).
DOI: 10.1093/oso/9780198911630.003.0006

Driven by Ramírez-Andreotta's aim to democratize all phases of science, one of the key questions for this work was how to best bring the data results back to the community for reflection, understanding, and action. She brought on multimedia artist Dorsey Kaufmann, who created *Ripple Effect*, an environmental art installation designed to tell the story of the water pollution data back to community members that collected the data, and to others.[2]

When you walk up to *Ripple Effect* it immediately evokes a sense of wonder and questioning. On a small transparent podium before you is a large black bowl holding water, with a ring of illuminating lights at its edge (Figure 5.1). You quickly notice that it is vibrating, causing drops of water to jump up and down on the surface. Kaufman worked with sound engineers to harness water *cymatics*—the visible physical response to sounds waves—and use the phenomenon to represent the data. One podium compared pollution levels against an agricultural irrigation standard, another showed risks of using the sample as potable water for humans, a third encoded impacts for livestock use of the water. Each played an audio track that was a transformed timelapse of the contamination levels detected in a single participant's submitted samples, creating more perturbations in the water via sound waves when their readings showed higher levels. The water droplets danced above the surface of water in each

Figure 5.1 Participants reacting to Ripple Effect.
Credit: Courtesy Dorsey Kaufmann.

168 OPEN MANY DOORS

bowl, highlighted by light-emitting diodes (LED)s that flashed on every time there was a chemical level that went above the regulatory standard represented at each podium.

When we spoke in an interview about the piece, Kaufmann described using art as a method for environmental data sharing. She was driven to connect people back to the experience of being in nature, which is tactile, multisensory, and evokes personal memories. Kaufmann was inspired by data sonification and ripple patterns from the water itself, the subject of the data collection. The design is an example of using a close material metaphor in a data sculpture; water is both the topic of the data and the material being used to display it. This has deep connections to the core impact of the project, echoing the complex relationship the residents have with water.

The piece is visually striking and draws the viewer in, immediately opening a door to the information that looks very different from a chart. However, Kaufmann and Ramírez-Andreotta wondered if it really worked or not as a communicative device for the more complex data story. They decided to study how people reacted to the art piece and the more standard printed booklet of data they also printed. How deep did each pull people into the story of the data? Was one more effective at conveying a sense of risk and concern? Did one or the other drive people to action?

The team constructed a controlled study comparing the booklet and the interactive piece; one group of residents saw just the printed booklet, while another was introduced to both the booklet and *Ripple Effect*.[3] They brought the exhibit back to the settings of data collection, inviting resident participants who had collected data samples to high school gyms, library basements, and other local community buildings to show them the story of the data. In these settings every resident participant was given small USB sticks, each holding their data readings about a contaminant of concern (lead, arsenic, and so on). They could walk up to a podium, plug their USB sticks into it, and see their data played on the water display. This immediately grounded the experience as one that represented the labor they had performed in producing the water samples. Like many sonifications, the compressed time axis of the playback showed them the full multi-season dataset quickly.

A first key finding from the study was that participants who experienced *Ripple Effect* expressed more intention to act, a sentiment that persisted even six months after the showcase (their methods reveal this finding as a significant correlation). Those who just received the booklet had fewer memories, and less specific ones, about the overall event when contacted six months after. Furthermore, their analysis of conversations around *Ripple Effect* showed more themes of fear and sadness in discussions of groups who saw the piece, emotions that their qualitative analysis connects to willingness to act and change. The conversations around the printed booklet were more passive even when the concentrations of contaminants were higher. They hypothesize that perhaps the report was a more familiar format for reviewing results and didn't engender as much of a sense of alarm. The group receiving both the booklet and seeing *Ripple Effect* showed sense-making behaviors that leveraged the two objects, often checking their booklet against the water display (and the opposite). Via multiple layers of reading, the piece scaffolded the participants from surprise into a deeper and more nuanced understanding of the data they had collected, and then gave them the opportunity to interrogate it more closely via the water motion and booklet together.

Ripple Effect created a new and impactful way to know the data that residents themselves had collected about rainwater contamination. Kaufman notes how they tried to create something different than "an exclusive space where only academics could feel comfortable." The data sharing events in locally trusted locations put the data into a familiar context for residents. The project goals required the team to reach outside the standard forms of charts and graphs; Ramírez-Andreotta's intention to democratize science pushed them to create a multi-sensory data exhibit. Kaufman described how the piece evoked memories of water, bringing people together in physical space to promote collective understanding and sharing history much like an effective interactive museum exhibit does. *Ripple Effect* leverages audiences' eyes and ears to bring them together around their data, and the associated evidence base about impacts points to the potential of creating multisensory data experiences for people who might not take to standard charts and graphs.

Multiple ways of knowing

You probably know someone who describes themselves as a visual learner or has personal experience with muscle memory in sports you play. People have lots of ways of learning and knowing things. Different fields of work accept and center different ways of learning and knowing. Data visualization as it is practiced now centers abstract ways of knowing information. Popular data suggests creating broader approaches to working with data and representing it, in order to reach diverse audiences that bring various ways of knowing to an experience. We must *open many doors* to support multiple ways of knowing.

170   OPEN MANY DOORS

Before I go too far down this path, it is worth pausing to consider what "multiple ways of knowing" even means. The phrase is dense and academic sounding, perhaps lending itself to far too many interpretations. Educators have a word to describe the study of how we know things—*epistemology*. People have defined various epistemologies, categorizing ways of knowing in a wide variety of ways. Related popular definitions include the theory of Multiple Intelligences (listing cognitive styles),[4] and the idea of learning styles (based on modes of perception).[5] No matter how you list them, people have different ways they feel comfortable learning and holding knowledge.

I've found a taxonomy from the field of participatory action research (created by Heron and Reason) to be particularly helpful to my thinking about an expanded toolbox for data representation. Their *extended epistemology* defines four ways of knowing.[6]

- Experiential: knowing based on the experiences of being in a place; sensory knowledge.
- Presentational: knowing from encounters with experiential knowledge, often through the presentation of music, movement, and so on; artistic knowledge.
- Propositional: knowing of something via ideas and theories that propose to represent it; experimentally informed knowledge.
- Practical: knowing via the performance of some action within some community of practice; applied knowledge.

These four present a helpful model within which to situate my argument about opening many doors to data. In their taxonomy we can see that current data visualization practice centers on propositional knowledge, the representation of some concrete thing via visual abstractions. In contrast, *Ripple Effect* offered experiential and presentational forms of knowledge.

Let me further concretize it with some illustrative examples. Reconsider Joiner's living histograms, which created a representation of the height of a group of people by having participants line up (experiential) instead of showing a bucketed bar chart (propositional) (see "Data theatre as embodied learning" in Chapter 4).[7] I think you can imagine how the physical experience of lining up leaves you with a very different understanding of the median height. Or take how I understood the prospect of becoming a parent by taking classes with other soon-to-be parents (practical), and also by watching movies about new parenting (presentational). We all participate in various ways of knowing, even if we don't name them. Multiple ways of knowing co-exist within one person.

Pulling harder on the thread of educational theory, another way to think about how we engage diverse learners stems from the idea of *epistemological pluralism*.[8] Proposed by Sherry Turkle and Seymour Papert, the concept emerged from their work to push the field of computer science to think more broadly about gender when considering the lack of women in the field from the 1980s on. Computer science education is built on right answers, introducing students to rigorous evaluation criteria based on the idea of computer programming as a practice of squeezing efficiencies out of computational hardware with limited resources. Most are taught that the "right" way to solve a problem in computer science is the most computationally efficient one, the solution requiring the least amount of calculations. Turkle and Papert instead argue that to engage learners more effectively we should engage and value the various ways they solve computational problems, sharing a rich set of evidentiary examples of students solving programming problems in "non-optimal" ways. One student "wants to manipulate computer language the way she works with words as she writes a poem." Another builds programs like she learns music, "perfecting the smallest little bits of pieces and then building up." Each is chided by their teacher for taking a non-standard, sub-optimal approach to creating their programs. Turkle and Papert argue that this idea of a "right way" is a root problem contributing to gender imbalance in computer science; the field needs to embrace multiple ways of knowing to overcome its challenges with gender diversity.

The ideas motivating their concept of epistemological pluralism are all around us and deeply embedded in our own experiences. Popular data focuses on settings where the dominant propositional approach creates friction and isn't aligned with underlying goals, turning some people away from engaging with data. When bringing a diverse group of people together around data we need to open more doors by embracing multiple ways of knowing. One key approach is by exploiting a broader range of sensory experiences. On the fringe of data storytelling practices, we can learn from ideas for hearing, smelling, and tasting our data. These open a variety of new doors, unlocking design potential for creating multisensory data experiences in community settings that allow for multiple ways of knowing data.

172   OPEN MANY DOORS

Hear your data

Click.
Click.
Click. Click. Click.
Click. Click. Click. Click. Click. Click. Click. Click. Click. Click. Click. Click.
Click.
   Click. Click. Click. Click. Click. Click. Click. Click. Click. Click. Click. Click.
Click.
   Click. Click. Click. Click.
Click.

These clicks hopefully suggest a device you've seen many times in dramatic movies: the Geiger counter. The hand-held tool detects radiation and represents it as sound. It works via a chamber of gas that gets ionized by radiation, creating a small electrical current that turns into sound (after amplification). This transformation is designed to allow the person holding it to concentrate on their potentially hazardous surroundings, instead of constantly monitoring a gauge or display. Outside of this well-known example it is unlikely you've run into any other representation of data as sound, yet there is active work on the idea in professional storytelling, artistic settings, academic research (and growing work in journalism).

   We can look beyond our sense of sight to find inspiration for approaches to opening doors into data via other senses. A first comes from a surprisingly long history of creating *sonifications*—data stories told through sound. The term has been in use in academic circles since the late 1990s, being more precisely defined in a 1999 NSF-commissioned report on the emerging field as "the use of nonspeech audio to convey information."[9]

   The last decade has seen major growth in broader application of sonification for data storytelling. Computational tools are being built to quickly experiment with mapping data from a spreadsheet onto pitch, rhythm, and other sonic attributes. Composers are creating performances about pressing social issues that represent data through their music. Academics are studying patterns of sonification to tease out the impacts, perceptions, and utility of turning your data into sounds and music.

   The first narrative sonification I ever remember hearing was on the radio. It was part of a larger British Broadcasting Corporation (BBC) report looking back at the North Atlantic Treaty Organization (NATO) bombing campaign in Bosnia and Herzegovina. To simplify a complex series of events, the Bosnian Serb army was accused of massacring civilians in supposed United Nations

(UN) safe areas, leading to an internationally sanctioned air-based bombing campaign to remove their strike capacity. In concert with the UN, NATO forces dropped over 1,000 bombs during August of 1995.

The piece I remember hearing was a compressed retelling of this bombing campaign. The journalists crafted an audio segment that played a sound of a bomb exploding for each time a bomb was dropped, essentially compressing the month of bombing to just a few minutes. It started off slowly, just a few "booms" playing, but then rapidly turned into an aggressive cacophony of sound. I remember feeling jarred, as if I had experienced some kind of conflict myself. The audio retelling pulled me in and left me emotionally shaken. This journalistic piece demonstrates a common approach to sonification and a frequent desired outcome, compressing events over time into a soundscape to create emotional impact on the listener.

More broadly, sonifications can take multiple approaches to representing data via sound. Computer scientists Thomas Hermann and Helge Ritter proposed a useful taxonomy to think through the possibilities of sonification in a 2000 academic paper. These included:

- *earcons*—abstract short sounds used to represent some event or information;
- *auditory icons*—short sounds designed to sound like the thing they represent;
- *audification*—sounds created by translating wave-based data through frequency-shifting and time-compression into audible ranges;
- *parameter mapping*—controlling attributes of sound like pitch and volume based on attributes of the data set;
- *model based*—sound landscapes that are dynamically changed through internal rules that respond to people interacting with them.

The BBC piece I remember used auditory icons—the sound of actual explosions—mapped onto a compressed time axis. A "timelapse" rendering of the data is a common element to many sonifications because sonic pieces play out over time, while visual depictions of data operate in two-dimensional (2D) space (*Ripple Effect* used the same technique). The previous taxonomy suggests the realm of possibility is much broader than literal representations of sounds (auditory icons). In fact, the longest history of sonification comes from a setting where it is used to represent something we otherwise can't perceive (via audification).

Learning from scientists

Scientists have been using audio displays of information to assist in their research and communication for decades. The International Community for Auditory Displays (ICAD) has convened gatherings for interested groups since the mid 1990s. Papers presented there have discussed uses of sonification in astronomy, financial analysis, cartography, radiology, seismology, computer interface design, spatial navigation, and more.

Of those academic disciplines, astronomy is the one where sonification has become most common.[10] Astronomers attribute this to some of the unique factors of sound, such as its multidimensional nature (pitch, tempo, volume, location, etc.) and the *cocktail party effect*—our human ability to pick out one distinct sound in a crowd of them. Astronomers have huge datasets, and importantly they capture phenomena outside of our perceptual abilities. Their observations include data about stellar bodies too far away to touch, too big to comprehend, that shine in spectrums of light we cannot see, exist in more dimensions than we can imagine, and make noise in frequencies we can't hear. Astronomers work with data representing real world events we can barely comprehend, constantly mapping from those imperceptible spaces to ones we can work with to answer their research questions. I think this predisposes astronomers to be more willing to work with their data in sound. Their source data can't be represented directly in any of our senses, so why not use sound?

Astronomer Wanda Diaz Merced gave a very personal explanation of her use of sonification in a 2016 TED Talk.[11] She studies gamma-ray bursts, one of those bands of light that lie far outside our ability to see. Early in Diaz Merced's career she would study gamma-ray burts using graphs that showed the bursts of light intensity over time. However, after an illness that led to the loss of her sight, plots like this were completely inaccessible to her. Returning to the table of numbers underneath, she created a way to continue her research using sound-based representations, mapping the intensity of light to pitch. Importantly, she shares in the talk how bass notes in the sonification revealed a pattern of resonance in electrical bursts that was hidden in the visual depiction. For Diaz Merced, the opening of a new door via the sonification not only let her continue doing her astronomy, but it also led to new insights. In Hermann and Ritter's categorization, her approach is audification, lending us another example to flesh out the space beyond the auditory icon-based BBC bombing report I remember so vividly. Merced advocates for using sound and visual depictions of data together to push astronomy forward.

The use of sonification in academic research is widespread beyond this intriguing and inspirational example from astronomy. Lengthy books have been edited about the theory of sonification, our perception of it, sonic interactions, design of sonification, use in the lab, application to industry and research, and more[12]. Compared to representing data via the other non-visual senses that I'll introduce next, ear-based data representations are quite well studied. However, they aren't widely used outside of this academic domain. This begs the question of why it is that only now sonification has started to spread more in the arts and other social sectors. Part of this can be attributed to the overall resurgence of datafication and the associated hype. However, there are also some real barriers to creating sonification.

A retrospective paper presented at the ICAD gathering in 2019 explored some key challenges to sonification that have slowed its spread.[13] A first challenge concerns precision—research has shown that we can be far more precise with our vision than with our hearing. A second challenge is related to individual perception—people have wildly different gut instincts about what changes in sound mean. While one person might think a higher frequency pitch means more quantity, another might think the opposite. Even more confoundingly, the various aspects of sound are perceived in inconsistent ways; louder sounds can sound higher pitched, speeding up tempo a certain amount isn't perceived the same way as slowing it down that same amount. Another challenge relates to the authors of sonifications—they frequently involve musicians as creators. Musicians perceive and connect to sound in different ways than non-musicians, sometimes creating explanatory sonifications that aren't easily understandable to other populations.

The paper that laid out these challenges was intriguingly titled "Is Sonification Doomed To Fail?" The author, Prof. John G. Neuhoff of The College of Wooster in Ohio, calls for splitting the field into two areas of research—artistic sonification and empirical sonification. Artistic approaches should focus on aesthetic perception, taking liberties with precise audio mappings of data in the service of the narrative experience the author is trying to create. Empirical approaches should build on existing knowledge to focus on mappings to audio that are easy to understand in precise and accurate ways for lay audiences who don't have much musical experience. The BBC piece on bombings is an example of this, utilizing a very simple and understandable mapping. Neuhoff argues that being more intentional about which space you are working in leaves muddy waters behind, offering helpful constraints for the appropriate design and evaluation of the impacts of your sonification.

I find this call for bifurcation quite compelling because it brings to mind challenges I've had creating both data sculptures and data mural. Sometimes data sculpture creators get stuck on trying to be empirically accurate when they are taking a more artistic approach. Similarly, sometimes data murals are questioned about where the data truly is being represented. This parallels questions about functional approaches to data visualization versus more artistic goals and leaves us with an open question about the level of detailed accuracy needed in encoding data in order to still count as data representation. The prompt has no simple answer, but I take Neuhoff's explanation to leave us with the reminder to be intentional about which approach we are taking.

Learning from musicians

Within the realm of artistic sonifications, there are groups already doing socially impactful work and opening new doors to data with sound. Since 2014, the California-based ClimateMusic Project has been creating musical pieces and hosting live performances that are based on climate data.[14] A piece titled *What If We ...?* from 2019 offers a good example of the type of parameter mapping sonification (per Hermann and Ritter's classifications) they have been producing. Describing sound in text is challenging, but I'll do my best here (you're best off finding a video and watching a bit to get a sense of the piece). Beginning with sea level rise estimates, composer Wendy Loomis and researcher Alison Marklein built an audio backdrop for their piece by compressing the time axis and mapping sea levels to audible frequencies; the higher the sea level rise, the higher the note that was playing. On top of this they included patterns of bass sounds representing land loss. These build in cacophony to challenge rhythmic drumming (representing sea level rise). To this they added recordings of hypothetical news reporters reading out imagined future headlines—"September 15, 2045: In the UK the Houses of Parliament and Westminster Abbey are Both Flooded." A traditional data chart showing sea level rise over time is projected to one side of the stage during the show, complementing their composition and performance.

The second half of the piece offers an alternative path that contrasts with this first, with less tumultuous music that is driven by a model of the world taking action to reduce sea level rise. You can hear the difference between the two paths; they've used the emotional impact and texture of music to tell two radically different data stories through sound. ClimateMusic Project has performed pieces like this over a dozen times around the world. After a piece finishes, they lead discussion with the audience and often partner with local activist organizations to create pathways to action. This mimics the approach I'm taking with data theatre, offering a pathway into civic deliberation after engaging a performative experience of getting to know the data (see "Paint data murals" in Chapter 3).

What If We ...? offers an alternate approach of science communication, one sorely needed to build support for acting on climate change outreach. The musical narrative appeals to a different way of knowing, in this case complementing traditional graphs in a way that makes the overall invitation to act more inviting and resonant with diverse audiences. The performance of the musical act itself connects to presentational ways of knowing. The shared act of listening triggers experiential ways of knowing the data that connect audiences to one another. One could close their eyes and simply let the experience wash over them, or pay more attention to the charts, narrative, and discussion to dig into more layers of reading. Science isn't normally taken to performance halls with guitars, drums, and singers; the ClimateMusic Project is meeting another set of people where they are, a group who might not otherwise engage with this data. They are opening new doors to understand complex impacts of climate change, evoking emotional responses via their sound-based explanations. The convening around music and post-event discussions rely heavily on well-known approaches to community organizing and engagement building regularly used by nonprofits.

Learning from digital tools

With modern multimedia technologies the power of creating a sonification isn't only in the hands of experienced composers and radio news narrative editors anymore, perhaps making Neuhoff's point about professional composers no longer relevant. The Google News Initiative in 2019 funded the development of a piece of open-source software they titled TwoTone to create more experiments in the space with new authors (Figure 5.2). The web based software, built by Datavized Technologies, makes it easy to map different columns of a data set to different attributes of music such as instrument and pitch; it lets you create pattern mapping sonifications in your web browser. After uploading a spreadsheet of quantitative data, you are dropped into a multi-track audio editing interface, familiar to anyone that has worked with GarageBand

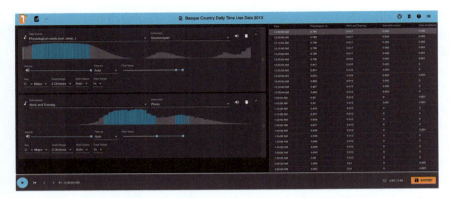

Figure 5.2 A screenshot of a multiple-track composition in the TwoTone desktop application.
Credit: author.

or similar tools. You select a column to sonify on that track, filter the input data to specific value ranges, and select the key, octave, arpeggio, scale range, and tempo. Other controls let you specify the total duration of the piece and how long each track is within it.

Through the combination of multiple tracks that represent different columns of filtered values, you can create very complex representations of the underlying column-based data. TwoTone makes it very easy to create a parameter mapped sonification. The designers' goal was to help reporters and others understand their data via sonification (like astronomers) and also to make musical pieces based on data (like the ClimateMusic Project). In fact, they cite astronomer Diaz Merced's TED Talk as inspiration for creating the software.[15]

Digital tools like TwoTone make it easy to convert data into sound, but don't offer any advice or guidance on if and when you should convert data into sound. Indeed, many examples of sonification seem to be playing with delight and whimsy at the expense of impact and responsibility.

With the widespread distribution of powerful multimedia devices in everyone's pockets and desktops, others have begun to explore sonification for data storytelling as well. TU Delft prof. Dr. Sara Lenzi and my Northeastern colleague Dr. Paolo Ciuccarelli decided in 2020 to catalog evocative examples by creating the online Sonification Archive.[16] The growing index now documents almost 500 digital examples, spanning data stories on dozens of topics created for various goals over the last few decades. That year they dug into a smaller set to understand a question relevant to the goals of the pro-social sector I'm focused on—how were sonifications being designed to intentionally focus on social issues? Their analysis included case studies with public audiences, not internal tools like most of the astronomy examples, and assessed the intentionality of the pieces to assess how they centered questions of social justice.[17] One of their key findings is that the process of designing a particular sonification needs to be more intentionally driven by the goals and audiences in pro-social settings. In many ways this is the same argument I make for the data visualization field writ large—the methods and media we choose to use when working with and representing data need to be aligned with the context of use.

Sonification and community

The artistic practice, research use, and professionalization of sonification show how quickly it is moving as a practice ready for the wider public. Interest in data sonification has grown alongside hype around data itself. Building from general approaches to auditory displays, various academic fields are asking questions about how to effectively map from data to sonic attributes. The answers are of course contextual, related to the goals and audiences of any particular sonification project. Sound has a long history creators pull from when crafting sonification, from music, to film, to video games, and beyond.

Like artists, news organizations have been leading the exploration of more public approaches to sonification. In 2010 the *New York Times* experimented with web-based sonification in a piece about just how close finishes are in the Olympics[18]. Various competitions are represented visually in rows, each with a play button next to them. Clicking that plays a series of closely spaced beeps, each indicating the finish of a person that didn't win the Gold; the beeps are spaced out in real time. Listening to the replay of just how close the finishes are reveals how tight the sports are, coming down to *Fractions of a Second* (the title of the piece). In 2019 a *Financial Times* explainer video argued "turning spreadsheets into sounds could be the next big thing."[19] In 2021 journalists Miriam Quick and Duncan Geere launched the *Loud Numbers* podcast to share experiments and lessons learnt about the potential for impact of representing data with sound.[20] There are more examples I could share, but this sample shows how data journalists are occasionally deploying sonification in their digital stories to aid in understanding, and how the field is reflecting on the potential impact and methods that best serve their needs.

A collaboration between Data4Change and civil society organization (CSOs) in Ethiopia offers another example of sonification in less academic

180 OPEN MANY DOORS

settings—the Hear the Blind Spot project. Based on survey data about digital exclusion within the Ethiopian visually impaired population, the piece included a sonification of results played on flute. Published both online and performed at an event, it was designed to push novel ideas of accessibility in the design of an experience for both sighted and vision impaired audiences, reflecting the inclusive focus of the data and report itself. Performed on stage with a flutist at the culmination of a multi-day event, local CSO member Awoke Dagnew described its inclusion as crucial to "build empathy and greater awareness."[21]

Sonification can use sound to bring the listener into another world in a way other approaches can't, offering narrative power to those trying to make an impact with data. Our ears are better than our eyes for finding patterns in time-based information.[22] The examples shared here leave me thinking that sonification is best fit for encouraging reflection and motivating action in groups of audiences that come together. Sonification is another door into data and can be effectively used to bring people together for building experiential knowledge collectively.

Smell your data

I have a strong suspicion that my wife is a super-smeller. She has an incredibly sensitive and finely tuned sense of smell. This helps in everyday settings, like telling me that my cookies in the oven are about to burn. However, it also extends to contexts I wouldn't have imagined; she can tell when someone in the family is fighting some kind of illness. This leads to sometimes awkward situations where she can smell that a colleague in a meeting is sick and should really be at home resting, but how do you tell someone that? After years of paying close attention to her quiet comments to me on her smell-related observations I can tell you that she's 100% right about these detections.

While we've never had her officially tested, the phenomenon isn't unique to her. *Hyperosmia* is the official name for having a heightened sense of smell. For some people it comes and goes based on other health changes, such as pregnancy or having Lyme disease.[23] For others, like my wife, that's just how they experience life (at least until losses connected to COVID-19). This is at times a blessing and at other times a curse.

Either way, living with a super-smeller has demonstrated to me first-hand how we habitually under value our sense of smell as a mechanism to learn

SMELL YOUR DATA   181

and communicate. In 2015 researchers at the Weizmann Institute of Science in Israel decided to study the phenomenon of *social chemo signaling* by observing something most of us do each day—shaking hands with another person.[24] To understand one potential purpose of the tradition of shaking hands, the team designed a study that secretly filmed people who were asked to shake other people's hands in a controlled setting. Their video recordings showed a surprising result; not only do people regularly sniff their hands, but after a handshake this behavior more than doubled. They added in covert odors to their study design and noticed an increase in the behavior, validating the hypothesis that the behavior was at least in part smell related. As an academic I call this subconscious somatic data collection—an example of how we use touch and smell to gather information about our bodies and those of others without even realizing it. Other research has validated my wife's abilities and the potential purpose of this unconscious sniffing tendency, demonstrating that we can smell if someone is fighting a sickness,[25] discern if they are afraid or worried,[26] and smell a variety of cues related to attraction[27] and reproduction.[28]

For all these demonstrated uses, smell is arguably the physical sense we understand the least. Researchers in various disciplines have created taxonomies of smell but have largely failed to come up with something as stable and useful as the tree of life in biology, or the periodic table in chemistry. We marvel at the amazing smell ability of a dog, but seldom seek to harness that potential capability in our own selves. Despite our regular use of the "sniff test" as demonstrated in the handshaking study, we still don't fully understand smell. This is a fascinating sensory playground to work within for data storytelling, and artists and researchers alike have begun to explore it.

A first challenge is what to call this idea of representing data by scents. Naming it a "smell based data representation" is a bit of a mouthfull. Scentification, olfactualizaton, odorification—none of those quite roll off the tongue either. Try saying "data smells" out loud to yourself and you'll understand why that one doesn't quite work either. I've decided to call these *smellification* here because it strikes me as the most evocatively descriptive of the idea amongst other options. In a data visualization conversation if you were to raise the idea of "smellification" I think most others would understand what you're talking about (perhaps with a smirk). With a name now in hand we can begin to map out the space of ideas and examples to understand the potential for creating smell-based experiences that communicate data. How might smell contribute to a larger toolbox for working with data in new ways that meet people where they are more effective than data visualizations?

Learning from research

Data visualization research is closely tied to academic departments of computer science, so it comes as little surprise that we can find several examples of computer-controlled smell-releasing devices being built to study perception and possibilities of smellification. Early work comes from a 2001 Massachusetts Institute of Technology (MIT) thesis by Joseph Kaye on his *Symbolic Olfactory Display*.[29] His pilot studies showed that scent is better suited for continuous data display, rather than discrete, and offered a theoretical framework of *olfactory icons* (much like earcons of the sonification field).

Almost two decades later a global research team decided to fully take on the idea of studying our "information olfactation" capabilities.[30] Just as we have a history of research into using visual channels to communicate abstract information via visual representation in charts and graphs (see "Problematic norms of data visualization" in Chapter 4), these researchers wanted to understand how people perceive information encoded in a scent-based display. Of course, that kind of a scent display doesn't really exist, so they began by building one. Their solution to support initial experimentation included a set of twenty-four bottled scents topped with ultrasonic diffusers, set behind a series of fans that could be turned on to direct the odors, all controlled by a computer that could also vary the heating and cooling of each scent as it was released. With this apparatus they could control which scent was being released, the amount of that scent, the temperature of the liquid that held the scent, and the speed of the wind it was delivered with—a capable basic scent-based display they called viScent. This let the team conduct a study to explore research questions about how much people could differentiate between smells and use that to discern quantities encoded into scents.[31]

The small pilot study with viScent revealed a few insights to build on. For example, it turns out that our ability to detect the intensity of a scent isn't very strong for helping discern values; when they mapped the quantity of something onto the intensity of smell, people could not differentiate between small changes in quantity. When they changed the actual scent based on some value of the data that worked quite well to discern between values; people could tell when one scent was used versus another. Varying the temperature of the liquid the scent was stored in also made a big difference. If you're going to encode data onto smell to support reasoning between differences, it seems like changing temperature and type of smell are your best options.

Other work around the same time dug into the questions of how to manage scents within the idea of an *olfactory interface*—a smell-based analog to

SMELL YOUR DATA   183

our touch interfaces and visual interfaces for computers. A team from the University of Sussex proposed approaches to delivering scents, monitoring their delivery and diffusion, and how to clean them out so they don't linger after needed. Their OSpace project created an enclosed room to focus scent-based experiences, a very similar scent delivery system to viScent, and a large roof-mounted fan to pull air out.[32] Not the kind of space I'd want to sit in for any extended period, but enough to perform very similar experiments on scent-based experiences and interfaces controlled by a computer.

Of course, not everyone has an olfactory display like these two lying around to work with, or the capacity to build one. In addition, this type of investigation into the space of possibility for smells and sensing and data doesn't really address what you'd want to do with one if you did have it. A study from computer scientists in 2014 surveyed over 400 people to try to understand what smell does for us.[33] They identified ten categories that offer a helpful framework for thinking about applying smells in ways that might make sense to people.

1. Smells associates us with the past.
2. Smells can quickly trigger memories.
3. Smells stimulate and delight us.
4. Smells create the desire to experience more (or less) of something.
5. Smells help us identify and detect objects and events.
6. Smells can overwhelm us.
7. Smells invade spaces.
8. Smells influence our social interactions.
9. Smells change moods and attitudes.
10. Smells build anticipation and change expectations.

This list presents one breakdown of the ways that smells already influence and interact with our experiences of the world, offering a handy idea of sensory norms we can take advantage of when thinking about smellification. Various projects reflect these roles that smell plays in our lives, offering more concrete inspirations for how smellification might function in community data settings.

Building on the role of scent as an ambient background sense, in 2017 German researchers created the inScent project. It consisted of a necklace, paired with your phone, that can release one of eight scents based on who was calling you.[34] They hoped to leverage how smells can help us identify some event and delight us as well. In their initial tests the triggers worked; people who wore the device could discern who was calling based on the scent released. Of course, this has some social problems: maybe not everyone around you wants to smell what was released. I'd certainly be thrown off if I was standing in a subway car and suddenly smelled a waft of rose scent from someone next to me who then pulled out their phone. However, is it that different from hearing their phone ring? Or am I just more accustomed to sounds coming from someone else's device? Their overall takeaway was that smell is well-suited for tasks like the notification one, where scent is used to reinforce and amplify some other signal.

Learning from arts and museum settings

In food industry circles, Boston is known as one of the origins of the American candy industry. Some of the key sugar-candy automation inventions have come from residents, including the slicing machine that led to Necco Wafers, invented by pharmacist Oliver Chase.[35] This history is still alive today, a fact I'm reminded of about halfway through my commute to work as I pass through Cambridge and bask in a luxurious sweet smell with traces of mint and chocolate. I go a little out of my way to make sure I pass the still-active factory that makes Junior Mints candies and savor the sensory moment every time.[36]

Smell is a strong trigger of memory. The smell of the Junior Mints factory as I pass it evokes feelings of my childhood, sneaking candies with my brother or sifting through piles after Halloween. Our smell pathways are tied to a variety of triggers within our brain. Projects from artists and exhibit designers are using that approach to experiment with data-related warnings delivered as smellification. Their examples show how scent might be used to convey information for individuals and groups of people outside of the lab.

Artists Leanne Wijnsma and Froukje Tan created a binary smellification called "Smell of Data" in 2017.[37] Their idea was that scent was critical for early humans to avoid danger—rotten food, nearby predators, and so on. In the digital age this manifests as new risks, such as the loss of our private data as we navigate the online world. One form of eavesdropping can happen on unsecured Wi-Fi networks, where attackers can easily see all the data coming in and out of your personal device. Wijnsma and Tan decided to bring this ancient role of smell as a warning into the digital age, creating a small device that would release a puff of scent to alert you when your device joined an unsecured network. Their piece intentionally used smell to remind us of the ancient conscious and unconscious patterns of using smell to protect ourselves. Here again we see odor being used to alert us to some event in our space, in this case being used to represent whether our personal data was being leaked or not. The piece helped reveal something we can't see, the invisible data flying around us on radio waves, data that might be revealing something private or risky about ourselves.

An earlier project from Olivia Alice and Kate McLean offers a more participatory and community-oriented example. They created a smell map of Milan presented at the annual Milan Design Week event (Figure 5.3).[38] Capturing the scents around the city—fish at the aquarium, tomato and basil at the restaurants, coffee at the cafes—they built a 3D table-sized model of the city with those smells embedded in the locations they were captured (via diffusers). The participatory project invited visitors to write what they thought the smells were on the buildings of the map itself. McLean has gone on to create similar smell maps of dozens of cities, amongst other scent-related artistic projects.[39] The map here used smell to create the desire for something more, offering a tantalizing quick scent of a part of the city most visitors had their own experience of and could connect to.

Outside of these arts projects, perhaps the most widespread use of smellification comes from the world of museums, who are using scents in a variety

Figure 5.3 A visitor smelling the "smell map" of Milan.
Credit: Olivia Alice and Kate McLean.

of ways to create more immersive experiences for visitors.[40] University College of London professors Cecilia Bembibre and Matija Strlič lay out a case for this in a 2017 paper that offers a framework for picking and sharing historic smells to evoke a sense of heritage.[41] They introduce how using smells in a museum setting can contribute to better learning, more personal connections to exhibits, and more overall enjoyment. The Jorvik Viking Centre in York, UK has for decades added pleasant and unpleasant smells of the era to their recreation of an ancient Viking village. A World War One museum in Belgium used the scents of chlorine gas and corpses to evoke the horrors of war. Various efforts have been made to produce smells of paintings, trying to evoke time and place with the data from the painting itself—plants, people, climate, and so on.[42] The National Museum of Singapore retold the country's history through a series of scents produced with a perfumery.[43] Museums are using smell to reinforce and augment visitor experiences, demonstrating how pushing visitors to intentionally pay attention to their sense of smell can engage them in new ways and represent information related to the exhibits they are interacting with or observing.

Smellification and community

Is smellification a technique that is useful in community settings? This sample of projects is very computation-heavy, but also includes some uses of smells in social settings such as museums and arts experiences. Most work across these spaces suggests that smells are best used in concert with other sensory cues but can be very helpful for differentiating between things. There are some important limitations that come up repeatedly, such as how our sense of smell rapidly adjusts to new norms. This phenomenon, called *olfactory overload*, is why a person in a kitchen baking cookie doesn't realize how lovely it smells, but someone walking in for the first time will immediately exclaim about the delicious odor.

Research from non-computational disciplines has shown that smells can be very effectively used in parallel with other sensory indications and evoke surprising connections within our brains. People consistently associate odors with features of hearing (such as pitch)[44] and seeing (such as geometric shapes).[45] This inspires the idea of linking the use of smell-based data representations to reinforce sonifications or data sculptures. Taking a multi-sensory approach to a data representation might help engage new groups of people in even traditional representations of data.

The core question for me is what perceptual and enabling elements smellification brings besides whimsy and delight. What doors does it open? The projects I've shared show examples of how smellification could pull broad groups of people into a story and evoke memories, but is the approach strong for building empathy with the contexts and subject of data, like data theatre? We have multiple points of evidence showing that people are delighted and surprised by smells, and that they are highly memorable. Perhaps these properties can be leveraged to connect with an audience quickly and effectively, drawing them into a data story that begins with smells, but then reveals more complexity layer by layer.

Challenges do exist that make working with smells difficult in community settings. Multi-scent smell machines, like those prototyped by computer scientists, are a long way off from being widely available, affordable, and easy to use. However, in the US a trip to your local natural goods or body and beauty store reveals a plethora of essential oils and other bottled scents to work with that aren't prohibitively expensive. Electronic scent diffusers are also widely available and quite affordable (marketed at people looking for home fragrance or relaxation solutions). These could be easily repurposed, like museums are doing, to create smellification experiences. While not data dense (in a Tuftean sense), such hacked-together solutions could reinforce and amplify data represented in multiple modalities, as noted by several of the research studies describing scent-based experiences. Most people want to stop and smell the flowers a lot more than they want to stop and read a bar chart, use that to your advantage by creating multi-sensory data experiences that incorporate scents.

Taste your data

The worst brownie I've ever had was given to me by a graduate student in a data storytelling class I was teaching at MIT. It was so salty that you could barely describe it as edible. I lunged for the glass of water offered and wondered how I had gotten myself into this situation. The salty aftertaste lingered all afternoon, well past the end of class.

Taste is the sense we first use to explore the world. Any parent will remember how their young child explores anything close to them with their tongue, leading to the parents experiencing strong feelings of both anxiety and amusement. Taste motivates and engages us throughout our lives, shaping our experience of the world, opening doors to new cultures, and bringing communities together.

188 OPEN MANY DOORS

Individual and cultural preferences polarize us around questions of sweetness, fermentation, and others. The power of food in community building is well-known, from pot-luck dinners to restaurants. Eating is part of our history, culture, economy, and of course a routine daily practice necessary for our existence.

Against this cultural and practical backdrop, taste presents a rich playground for data encoding. Taste is driven by multiple attributes of texture, temperature, flavor, and of the visual presentation of food as well. Some are not that widely known, such as the critical role of smell in taste, or the fact that spiciness is really pain. Others are broadly enjoyed, such as the combination of savory and sweet, or crispiness and crunch. In addition to personal proclivities, our taste is driven by evolutionarily important cues—sweetness implies energy density, bitterness is a warning signal of possible toxins.

Taste is complex.

The over-salted brownie I had at MIT was not complex. It was just plain terrible, but it was mostly my own fault that I had to eat it. Near the close of that semester's course, I had asked the students to create edible data representations. That group of students decided to dive into air quality data. That type of data is hard to experience; we don't see subtle changes and don't have a good sense of what an abstract number on air quality hazard scale means in terms of our daily experience. Yet it has become increasingly important as forest fires and other natural events driven by climate change create far more hazardous air condition events and regular warnings about going outside.

My students choose to represent the experience of hazardous air quality levels via brownies. Gathering data about air quality in Los Angeles and Beijing, and official standards, they created three brownies where the amount of salt was determined by levels of small particulate air pollution in 2014 (Figure 5.4). The first brownie was labeled in green and represented the World Health Organization's standard for safe air quality. It was quite delicious; smooth, and gooey with just a hint of salt to bring the flavor out. The second was labeled yellow and represented the worst level in Los Angeles that year. It was clearly over-salted but still edible. The final was labeled in red and represented the worst day of air quality in Beijing that year. To be honest I almost spit it out, but trying to be a committed teacher I swallowed it down and followed it with several chugs of water (much to the amusement of the rest of the class). Teaching edible data has its hazards.

Over multiple years I've found that students in my journalism classrooms gravitate toward ideas of food-based data storytelling for news sharing. Some

Figure 5.4 Students baked brownies with varying levels of salt based on air pollution data.
Credit: author.

work with simple literal encodings, such as representing honey production volumes via jars filled to various levels. Others map data values onto eating experiences, like the brownies baked to represent air pollution levels in different cities. Still others play with the performative aspect of tasting their concoctions; one team creating smoothies representing countries, with the amount of spinach based on per-capita greenhouse gas emissions, and then filming their reactions to tasting them like a late-night TV show bit. These examples suggest to me that data learners are well equipped to explore food-based encodings and find them engaging for audiences that might not otherwise be interested in the topic the data is about. Everyone connects to food and taste; it meets people where they are in obvious ways.

Learning from research

My journalism student's projects are examples of *edibilization*—varying some taste-connected aspect of an object you are intended to eat based on data. The brownie-baking students choose to vary the saltiness based on air quality. They could have also change texture, temperature, shape, or plating (those last two are less unique to edibilization so I won't engage them deeply here).

Thinking through how to vary taste can be complex. Chefs designing dishes work with body, senses, location, climate, memory, economics, performance, cultural history, and more. Sometimes their design choices are active and intentional, such as working with seasonal ingredients based on a desire to buy local. At other times choices are passive or embedded in other decisions,

190 OPEN MANY DOORS

such as working with the cheapest ingredients in the market based on a desire to keep costs low. There is a plethora of writing on food and culture, restaurant experience design, and more that might have relevance for approaching edible data. Your setting impacts how things taste and how you enjoy them. Your emotions have similar impacts. The eating experience is nuanced, especially when connected to technologies or other aspects.[46]

A 2016 paper from a team at the Hong Kong University of Science and Technology introduced the idea of data edibilization to explore how a broader range of senses could be used to experience data stories.[47] To study the potential in comparison to existing approaches, they created three edible data dishes and invited about a dozen people to eat them and compare them to a graph of the same data. Across the board people found the edibilizations enticing, in no small part because the study was hosted just before lunch and participants were starting to feel hungry. Participants also noted the visual presentation of the edibilizations were more pleasing and more fun than the chart. The study included evidence in line with our existing understanding of the links that make taste a strong trigger for memory, much like smell.[48] The multisensory experience of the edibilizations stuck with participants. On the other hand, the legibility of the edible data dishes was not very highly rated; participants in their small study found the details of the data in the dishes harder to understand than in the charts next to them.

What is the potential power of edible data representation? One clue can be found in a side note from this study. Intriguingly, once participants understood one encoding, they began to wonder if more existed. It appears the idea of size-based encodings stimulated questions about flavor-based encodings. More interestingly, many commented on the flavor choices once they began to reflect on them, speaking to opportunities for layers of reading. For example, a cheese-based dish that represented an Asian-American description of themself encoded responses onto colors and flavors of cheeses, leading one participant to note:

I notice that the texture and flavor of the white and mixed-color cheeses are more similar compared to the orange one. Maybe it means American and bicultural Asians can blend into the U.S. society better than those who maintain their ethnic identity?

While the visual presentation of an edible data dish might be appealing and allow for one layer of understanding, the taste might carry another. The participants were jumping very quickly to questions about multiple encodings of data onto flavor, color, and more. This experimental result suggests to me that there is an opportunity here for edibilization to be more evocative and explanatory than simply visual.

Learning from culinary design

Beyond the classroom and the academic research lab, design and cuisine experts have also explored the idea of data and food in other settings. The most well-known project is the series of Data Cuisine workshops launched in 2011 by data visualization expert Mortiz Stefaner and curator Dr. Susanne Jaschko. They suggested the notion of *culinary variables* to help think this through—an idea that data could be encoded onto flavor in discernible ways.

We can tease out how this looked by digging into one piece: the *Pizz'age* dish created by data chefs Léa Johnson and Alyse Yilmaz (Figure 5.5). They chose to work with life expectancy data from three countries—Sierra Leone, France, and Japan. Each country was represented as a rectangular pizza. The life expectancy for each determined the length of the rectangle-shaped pizzas and the amount of time each was baked for. The toppings were based on the cultural cuisine of the country; Sierra Leone included Harissa sauce and tomato, France had bacon, olives, and cheese, Japan had noodles and tuna. The overall dish was built around a metaphor of food as life-giving, while the

Figure 5.5 Pizz'age by Léa Johnson and Alyse Yilmaz.
Credit: Courtesy Mortiz Stefaner and Susanne Jaschko.

visual evoked the idea of a bar chart due to the length-based encoding when viewed from above. The toppings were a flavor-based encoding of the country, while the baking time was used as a way to make the preparation method a redundant mapping of life expectancy. The Sierra Leone pizza is under-baked, an apt metaphor for a life cut too short.

Looking across the set of edible data dishes from two workshops, and analyzing them for patterns in how they were using cuisine to represent data, reveals some of the potential of data cuisine. The 2017 workshop in Pristina, Kosovo and 2019 workshop in Paris, France offered a variety of edible data dishes to analyze (fourteen in all). A first question that surfaced was to what extent they employed existing visual data encoding approaches, and how often? This returns us to the questions of changing shape, size, texture, color, or position based on some data attributes. *Pizz'age* used the visual of a bar chart, mapping life expectancy onto length. Every single one of the dishes used at least one of these visual approaches to represent data. Size was varied to represent data most often (as in *Pizz'age*), then text and color in about half, with position only used in two.

What about non-visual sensory channels related to food? In this small set of examples flavor and texture are used to represent data, but also nutritional value, preparation method, and spice-level. *Pizz'age* changed the baking time, varying the preparation method to represent the life expectancy. It also changed the flavor in each of the three parts of the dish, using it to represent the culinary heritage of the country it represented. Almost every single one of the dishes varied the flavor of elements of the dish based on data like this. A third of them each varied nutritional value and the preparation method. Spice level was changed to represent data in just two dishes.

The Data Cuisine workshops offer a rich set of examples, and opportunities for inspiration. The workshops themselves offered a fun and evocative design experience, mostly couched in the language and practice of professionalized data visualization and museum curation. The participants engaged data through cooking and eating, demonstrating the potential of data edibilization as both a participatory tool and approach to building experiential knowledge.

The *Taste of Data* series created by Austrian designers Veronica Krenn and Veselea Mihaylova flesh out some more ideas of what edibilization can be, and for what purpose. The pair share one of my motivations, "to create a multisensory approach towards data,"[49] and instincts about the power of food, that it "can communicate emotions and memories, and this connects the audience strongly to the message sent."[50] They have created several edible pieces where the ingredients and food styling vary based on data. One series focused on the iconic sausages of the Eastern European region they are from. Responding to a 2013 local meat production scandal, which found sausage tainted with horse meat, sea food, and even toilet paper, they created sausages for different countries that mixed in tainted ingredients based on corruption levels in each country.[51] Based on the Corruption Perceptions Index report for 2012, the Bulgarian sausage they made contained 59% toilet paper.[52] The Austrian one had 31% horse meat and tofu. The repurposing of contemporary events, the metaphor of corruption of food, and the civic-related data all turn their performative design into a data-informed statement about corruption. This uses the power of food as a metaphor and a way to pique the interest of a public that might otherwise not turn to this data, even if it resonates with their lived experiences of corruption. These weren't intended to be eaten, because of the tainted ingredients, but were showcased via a video documenting their creation as well. Edible data can engage presentational knowledge, just as data sculptures viewed on TV work differently than charts.

The data chefs in Stefaner and Jaschko's workshops used traditional visual chart approaches in food form often, but also changed the flavor to represent data the majority of time. Krenn and Mihaylova played with food metaphors to represent food-related data. Both examples again speak to the potential of multi-sensory experiences, here combining taste, smell, and sight in novel ways to create food-based data stories that open doors for alternate ways to bring people together around data.

Edibilization and community

Across academic research, artistic interventions, and my own teaching I see examples of people exploring the idea of edible data. Like smellification, there is a strong human desire to experience flavors and dishes. These types of sensory experiences create very compelling first layers of reading, a very accessible and inviting way into a data-centered conversation in pro-social settings. Food has long brought people together, and cultural practices center food and eating experiences across the globe. This rich cultural backdrop to food means that edible data lends itself to being participatory and meeting people where they are. There is a power in using a well-known medium to engage people; everyone just understands food and is intrigued by it. Food entices. Food seduces. Food piques curiosity. Edibilization can make data easier to digest.

However, it isn't clear to me yet that there is depth of data representation in edible data. The early research shows that people have a hard time discerning attributes such as the intensity of a flavor. Like smellification, perhaps there is more value in comparing different flavors, or creating extremes of the same flavor. That certainly worked for the salty brownies my students created, which have lingered in my memory and taste buds ever since. No single technique of our expanded toolbox is right for every intended audience and goal. My intent in exploring these non-visual sensory-focused examples is to expand the set of options so that we can more appropriately match the forms we choose to use with data to the context and goals of use.

"Show up where they are"

Continuing the theme of food in data visualization, we come now to the avocado. You might be surprised to hear that it holds a vaulted place in the field. I've heard far too many times, at one academic conference after another, how avocados make the invisible state of ripeness of their delicious interior visible via the color of the outside skin. If visualization makes the invisible visible, then avocados are a great metaphor (the speakers' claim).

I've long found this confusing. I was taught to gently squeeze an avocado to determine its state of ripeness: using touch not sight. Later in life YouTube taught me to knock on a watermelon with my knuckle to tell if it is ripe: using sound not sight. I stick my nose to the bottom of a pineapple to check its ripeness: using smell not sight. I shop for fruits and vegetables with all my senses; going to the market is a multi-sensory experience.

This fraught avocado metaphor shows how we've privileged the visual in data storytelling in a variety of ways. It is time to engage the other senses and explore what they have to offer, especially for those looking to broaden their toolbox to use more effective methods of bringing people together around data in pro-social settings.

Data stories can engage senses beyond the visual in a variety of ways. The key question for communities is when to pull sonification, smellification, or edibilization out of your toolbox? Each has its own strengths and weaknesses, and hopefully in the years to come we'll continue to see more evocative examples being created to tease out new possibilities. Many of the research studies I've mentioned show us how our senses overlap, suggesting that creating multisensory data experiences can effectively reinforce narratives via opening multiple doors for different types of learners.

A 2012 segment from a morning TV news show offers a helpful example of how multiple senses can be combined to draw audiences into a data story.

Tony Dokoupil, co-host of the CBS *This Morning* daily live TV news and entertainment program, hosted a five-minute video segment that explored people's perceptions of American inequality in a novel way: with pie (Figure 5.6). The video opens with him trying unsuccessfully to get visitors to stop and talk about wealth inequality, until he starts using the visual explanatory metaphor of who gets a larger "piece of the pie."[53] The key invitation to participate is a prompt to take a full pie, representing all the wealth in the US, and distribute it onto five plates as accurately as possible, each representing an economic quintile: from the 20% most wealthy to the 20% least. The next few minutes show people pondering, cutting the pie, and moving it around the plates, trying to map out their guess at the distribution. The segment pivots when he repeats that "no one was even close;" almost all the pie should go on the "top 20%" plate, while the upper-middle and middle class share most of the last 10%. The rest just get crumbs, with the lowest quintile cleverly receive a bill for "pie debt."

The metaphor of pie is evocative, thanks in no small part to the well-known phrase about getting a "piece of the pie." Dokoupil closes by linking to contemporary narratives from politicians such as Elizabeth Warren and Bernie Sanders to return tax levels on the wealthiest to levels seen in the early 1900s. His participants universally think the disparity is a problem, advocating redistribution and grabbing their forks to enjoy "rich people pie." Participants in

Figure 5.6 A scene from the CBS *This Morning* show segment on wealth inequality.
Credit: CBS News Archives.

196 OPEN MANY DOORS

the short bit are constructing their understanding of the dataset out loud with the host, using their body's physical motions to interact with the dataset visually and physically by cutting the pie, and then producing an interactive data sculpture that represents their mental model of the dataset. This process engages both experiential and presentational ways of knowing, pivoting off the propositional knowledge of a standard visual pie chart. Experienced through the TV, this functions like a data performance for the broader set of viewers. Playful data experiences like this CBS TV clip can seem simple and shallow, but in fact carry significant design challenges and opportunity for impact.

Dokoupil's narration and interaction demonstrate how creative data representation can be used to engage complex and sometimes boring topics with the public; he's at a mall engaging people just walking around shopping. The data experience leverages the power of food and participation to engage people. The reveal grabs people by showing them how their perception of the data is incorrect, much like the participatory *New York Times*'s "You Draw It" graphs (see "Participatory data graphs" in Chapter 2). The impact is increased support amongst participants for policies distributing wealth more equitably. "It's disturbing" commented one participant.

People understand and engage with knowledge in multiple ways. Propositional ways of knowing (theoretical engagement with an idea via abstract representation) are the most common taken in data visualization. This is the knowing a scientist typically engages in when representing data. Centering this one approach excludes others in ways that can disengage audiences. Appealing to a broader range of senses can open doors for people that bring skills and wisdom that don't mesh with propositional forms of working with and knowing data. Being in a multi-sensory space with data that is more than simply visual allows for experiential knowledge to be built. Being an audience member taking in a theatrical or musical presentation of data allows for presentational knowledge to be built. Engaging hands-on with data amongst a community activity allows for practical knowledge to be built.

MIT professor and civic designer Cesar McDowell, when defining his idea of participatory *Big Democracy*, sums up his driving motivation in a way that connects to arguments I've put on the table here.

In democracy, in order to solicit the knowledge that the public has, you have to ask people in the form they're used to having conversations in, and you have to show up where they are.[54]

This is critical advice: "show up where they are." Public spaces are the location where community is fed and nurtured, and public events are a standard way to connect in those spaces. Arts practices have long known and leveraged public space in the service of contextualizing issues of importance. This can be traced back to the early 1900s, when artists across media began to exercise more control over the space where their work was seen; consider the difference between a painting in a gallery and a mural around the corner from your home to understand this point.[55] Recent research has shown how public art can be an economic driver and combat negative social issues in cities.[56] It is now widely understood that arts-based interventions significantly contribute to building and maintaining a sense of place.[57] The arts have a long practice of meeting people where they are to create evocative experiences that provoke new ways of thinking.[58]

We need to use all the senses to invite populations that are otherwise left out of data-centered processes that impact their lives. Data murals push at the edges of visual representations of data. Data sculptures explore physical and tactile representations. Data theatre offers immersive dramatic experiences of data. Data sonification engages via the audio processing and emotional cores of our brain. Data edibilization and smellification evoke memory, culture, and context in ways that can pull broader groups of people into data stories.

Multi-sensory approaches are applicable to all the pro-social settings I've focused on. Digital journalists can push beyond the visual to engage sound, taking full advantage of multimedia computational devices. Local journalists can create mixed sensory real-world data experiences with readers in the physical spaces where they are trusted. CSOs can build outreach programs and inclusive data events about the impacts they are having, inviting diverse populations to engage with data that helps explain the issues they work on. Science museums can design multi-sensory exhibits that draw in visitors and explain complex phenomena via data exhibits that offer many different sensory channels for engagement. Libraries can host performances that bring data about ongoing local issues to life via live sonification. The options are rich and diverse, presenting many opportunities to rethink who gets to work with data and how, if we just break out of our focus on the visual.

This collection of sublime, scientific, and wacky examples gives me hope that the tyranny of the 2D in data experiences is coming to an end. That's good news for anyone working with data in community contexts for the public good. Engaging the senses more fully can offer playful, inviting, and meaningful new approaches to bringing people together around data. Experiences that cross the senses trigger emotional responses and ways of thinking that can produce novel insights.[59] Our multisensory world has so much more to offer than flat graphics for empowering people with information.

Conclusion

Practice Popular Data

Popular data is a framework that offers a larger toolbox for bringing people together around data, specifically designed for pro-social settings in communities. This book catalogs the motivation and practice, building on inspirations from libraries, newspapers, museums, civil society organizations (CSOs), and other civic contexts. The work of activists fighting for housing rights in Los Angeles offers a final case study demonstrating the key principles in practice in the design of a data mini-golf course, satirical play, and interactive physical museum exhibit. Popular data will help you, like them, create impactful data experiences for community settings in novel ways.

Playful data in Los Angeles

Since the 1980s, Los Angeles' Skid Row has been thought of as an example of urban blight, regularly home to around ten thousand unhoused people, the formerly unhoused, and others. Numerous rounds of conflict around city policies have created a strong and supportive community fighting to maintain their place in the city. As urban planning has become more data centered, local CSOs have adapted, using the language of data to fight for homes in the constantly changing city. One particular group, the Los Angeles Poverty Department (LAPD),[1] has built an evolving practice of designing and creating data experiences that reflect the key principles and practices of popular data. Three of their playful and engaging interventions offer vivid examples of popular data in practice: a playable nine-hole data mini-golf course, an accompanying satirical play about corruption that was performed at the course, and a hands-on museum exhibit that lets visitors construct their own solution to the funds needed to create enough housing.

To understand the LAPD's work, we can start with a little background on Skid Row itself. In the 1800s Los Angeles was the literal last stop on the new

Community Data. Rahul Bhargava, Oxford University Press. © Rahul Bhargava (2024).
DOI: 10.1093/oso/9780198911630.003.0007

200 CONCLUSION

cross-country rail line. The area attracted travelers moving through, people looking to begin anew, and a significant amount of seasonal labor. Local hotels, bars, brothels, and other institutions of "ill repute" catered to their needs at affordable rates. The city grew and changed, but Skid Row retained a reputation as a place for those on their way from one place to another. A hundred years later, in the 1950s and 1960s, the city looked at the declining state of the built infrastructure in the area and pushed a policy of clearing damaged buildings, making it cheaper for property owners to demolish than to repair. This led to a significant decrease in housing units, increasing the population of unhoused people. The 1970s saw city administration adopt a policy of "containment," trying to concentrate the unhoused in one area (Skid Row). The 1980s saw a back-and-forth tussle between parties in government on whether to arrest the unhoused or provide more services. Residents were displaced, often by police actions.

Today Skid Row is still home to many unhoused, veterans, artists, and others. This mix of groups has created and cultivated numerous now well-established collectives and support groups to organize, advocate, and care for each other. The LAPD is one of those nodes in the network of support and community. Founded in the mid-1980s by artists and activist John Malpede, the LAPD facilitates a performance group made up of unhoused and formerly unhoused residents, activists, and other community members. It is known as the first theater group created for and with an unhoused population. Their performances reflect the social and political issues faced by residents, using theatre to highlight social injustice. Hosted at the Skid Row History Museum and Archive, the LAPD calendar of events offers regular theater workshops, open-mic nights, exhibits, film screenings, and more.

Around 2015 the LAPD started a new campaign of organizing, in response to new city zoning policy changes that threatened to remove large swaths of affordable housing and open it up to market rate (i.e., expensive) development. The plan, part of a larger rewriting of the zoning ordinances, was named "Recode LA" and was widely projected by activists to displace the current low-income residents of Skid Row.[2] The response from activists led to the projects I'm showcasing here as powerful inspirations for context-appropriate and effective creative data storytelling for pro-social community settings: *The Back 9* and *How to House 7,000 People in Skid Row*. These projects are strong examples of an alternative take on who gets to work with data, in service of whose goals, and how data is represented.

The Back 9: Golf and zoning policy in Los Angeles

My family knows one thing about vacationing with me: if we're anywhere near a mini-golf course, we'll have to stop and play a round. Created by an American entrepreneur during the expansion of the highway network, mini-golf holds a nostalgic charm for many in the US (not just me).[3] It is associated with family summer trips, wholesome fun, beaches, ice-cream, and joyful celebrations of a hole-in-one. A mini-golf course is generally made up of nine or eighteen holes, played with a putter and a standard golf ball. The key part is that each course has obstacles that players must putt around to sink the ball in the hole. Many people call it "putt-putt," which was the name of the company that produced kits and plans during the 1950s and oversaw its massive growth.[4] Courses usually have a campy theme—windmills, pirates, safari, aliens, indoor black-light effects. More recently I've seen more pop-up temporary mini-golf courses created to enliven special events such as museum openings (where holes mimicked the artwork), and temporary exhibits (such as the *Putting Green* in New York City which has the theme of challenges and responses to climate change).[5]

Inspired by this history of mini-golf as a popular playful pastime, in 2017 the LAPD decided to argue against projected negative impacts of the Recode LA plan by designing a history- and data-inspired playable mini-golf course they called *The Back 9: Golf and Zoning Policy in Los Angeles* (*The Back 9*) (Figure C.1). In an interview, project directors Henriëtte Brouwers, John Malpede, and artist Rosten Woo described to me how they designed *The Back 9* to "make the incomprehensible compelling." This team used the whimsy, playfulness, and approachability of mini-golf to pull local audiences into the debates and motivate action, "interrogating the power structures that have literally built Los Angeles."

Each hole was connected to related geographic, planning, and housing history and data. One was a wall-sized line chart contrasting rapidly decreasing housing capacity and slowly growing population. To play the hole you climbed the rungs of a ladder labeling the vertical axis, and rolled your ball down the hollow tube that represented the housing decreasing over time (Figure C.2).

The final hole was a conventional map of Los Angeles depicted on a ramp you had to putt up. The proposed new area for Skid Row was cut out, leaving a hole you needed to sink your ball down to finish the course. People would try to aim for it, but consistently failed to get the ball to go in. The "aha"

Figure C.1 Los Angeles residents playing The Back 9.
Credit: Courtesy of the Los Angeles Poverty Department, Skid Row History Museum and Archive.

moment came when they realized the space cut out just wasn't big enough to sink the ball through; a metaphor LAPD had created to drive home the message that there wasn't enough space in the new plan for their community to fit.

Through out the multi-month run of the mini-golf course LAPD offered public workshops in the space to explain zoning, public housing, and related issues. Some topics included "What is Affordable Housing?," "Know Your Property Rights!," and "What is Zoning?." These delved further into the context, history, and data explored on the course itself.

To offer yet another critique of the proposed zoning changes, Malpede directed the performance troop in devising a play about the back room wheeling and dealing that led to the Recode LA plan. Malpede shared with me that this was driven by LAPD's belief in "theater as a special space" that makes "certain statements sayable and ideas thinkable." The troop improvised dialog based on political leaders' actual statements, highlighting the legal collusion that led to the proposed plan. The satirical play was performed on the mini-golf course itself, reacting to the data it depicted while simultaneously mocking the idea of making out-of-sight political deals on real golf

Figure C.2 A playable line chart from The Back 9 mini-golf course showing the narrowing housing gap.
Credit: courtesy of the Los Angeles Poverty Department, Skid Row History Museum and Archive.

courses (Figure C.3). The satire was well-warranted, three of the political leaders being mocked in the play are now in prison due to their related illegal activities.

How to house 7,000 people in Skid Row

The organizing around *The Back 9* led, in part, to the creation of a new neighborhood coalition called Skid Row Now. This group amplified collective efforts, providing a vehicle to directly engage with developers and city planners. They created the "Skid Row Now and 2040" plan as an alternative, proposing changes that would lead to no displacement of current residents nor the unhoused. Going even further, the plan called for 7,000 additional units to be created, enough to house the number of residents that were in transitional programs or living on the streets at the time. Based on the report, and the predictions, the team created the *How to House 7,000 People in Skid Row* exhibit (*How to House*), which allowed visitors to physically craft their own

204   CONCLUSION

Figure C.3 A scene from the play being performed in front of an audience.
Credit: courtesy of the Los Angeles Poverty Department, Skid Row History Museum and Archive.

plan to pay for the housing needed. Their goal was to make the plan legible to a larger group of people, and they did it by employing a physical invitation in a museum setting.

The exhibit opens with a table full of plastic miniature people, giving embodied form to the 7,000 people their alternative plan proposed to house. Woo described the bulk of the exhibit as "walking around in a big math problem." After that first table, the visitor is then introduced to a volumetric encoding of the estimated cost to house the 7,000 people in need ($3.5 billion dollars), a physical box roughly two feet on each side.

As you continue down the central wall of the exhibit you begin to encounter potential solutions with their own boxes of varying sizes (Figure C.4). One shows the private real estate development pipeline (six feet tall). Another represents the funds that could be developed from a proposed city plan (twice that amount). Even more colored boxes show the sizes of other potential approaches such as inclusionary zoning, the equivalent value of existing vacant housing that could be taxed, new public housing investments, and other ideas. These are giving physical form to data about the variety of options that exist in the report published by Skid Row Now.

Figure C.4 The "How to House 7,000 People in Skid Row" exhibit.
Credit: courtesy of the Los Angeles Poverty Department, Skid Row History Museum and Archive.

With these proposals rendered in physical form, you're invited to mix and match solutions by stacking the boxes to reach the height needed. You might stack the inclusionary zoning box and a box representing a small new tax to get close to the amount (Figure C.5). The boxes are lightweight, making it easy to move them around and compare, much like the street-side cardboard bar chart created by Black Youth Project 100 (BYP100) in Chicago (see "Data sculptures as mirrors" in Chapter 3). The goal of the physical puzzle was to combine taxes, housing stock, and other proposals to get to the level of financing needed to create 7,000 housing units. Pushing our imaginations, the biggest of all boxes showed the need of the estimated 60,000 unhoused people in all Los Angeles.

The principles of popular data in practice

Communications scholar Aristea Fotopoulou argues that centering data projects with social impact on interactive experiences can create long-lasting benefits for participants such as skill development, empowerment, and

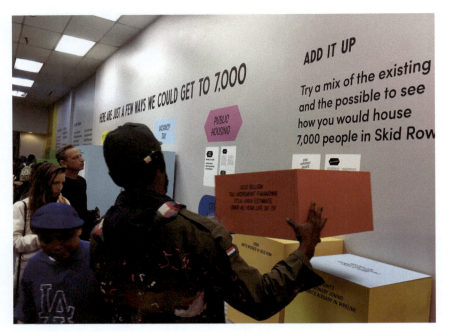

Figure C.5 Visitors trying to "add it up" at the exhibit by moving boxes representing costs and housing units.
Credit: courtesy of the Los Angeles Poverty Department, Skid Row History Museum and Archive.

inclusion.[6] Rather than having those outcomes as secondary, she posits that they are central in pro-social spaces, and the design of projects should engage them fully. LAPD put this philosophy into action with their work on *The Back 9* and *How to House*, creating concrete examples of the application of each of the five principles of popular data.

Both were clearly designed to *center impact*; they hoped to align public support with their housing needs and influence zoning policy to protect their community's existence. To understand how, we can leverage the social-ecological model often used in the domain of public health for health promotion.[7] This model links impacts at the individual, interpersonal, community, policy, and cultural levels (Figure C.6).

At the individual level we can consider the impact that comes from physically playing the mini-golf course or stacking the boxes. My summary of embodied learning research (see "Data theatre as embodied learning" in Chapter 4) shows how that activates different ways of knowing than a report.

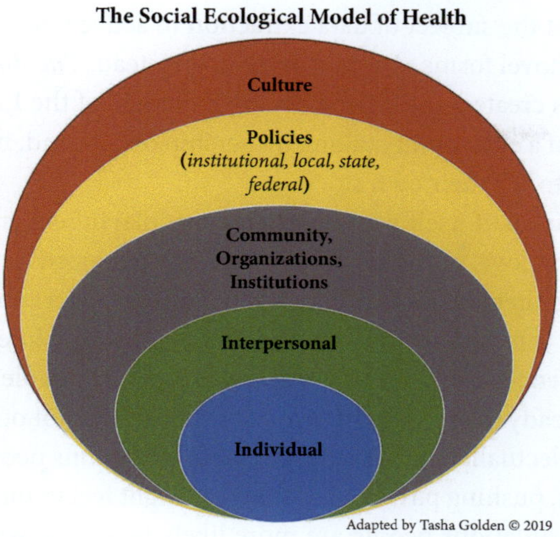

The Social Ecological Model of Health

Adapted by Tasha Golden © 2019

Figure C.6 McLeroy et al.'s social-ecological model of health, as illustrated by Tasha Golden.

Credit: CC 4.0 - Tasha Golden

At the interpersonal level we see how each interaction was designed to support both an individual experience and a spectator's, they sparked conversation between visitors because each participant was also performing the interaction for those around them. In both experiences, solution finding required working together. The play and workshops added to this, creating opportunities for construction of presentational knowledge interpersonally.

At the community level, the shared experience and space of both pieces cultivated a sense of belonging among residents and normalized discussion of problems and solutions based on public data.

The experiences were directly designed to impact policy. In fact, after visiting *The Back 9*, city employees on the relevant task forces built a relationship with the LAPD, leading to surprising examples of collaboration and support with the city itself.

The group itself is trying to change culture in their city around the unhoused and socio-economic disparities. In addition, they function as a cultural institution, curating and disseminating ideas like other museums and libraries do.

LAPD's work on *The Back 9* and *How to House* projects echo prior examples I've described that *focus on participation*. The group itself is made up of housed and unhoused residents of Skid Row, offering new ways for a group

208   CONCLUSION

that is usually just the subject of data extraction to actively participate in data production and novel forms of data storytelling instead. *The Back 9* theatrical performance was created and performed by members of the LAPD. This is a strong example of a community taking ownership of data and, in concert with experts, using it to tell their own story.

The Back 9 mini-golf itself was a participatory playful experience. Forgive me if some of my love of the pastime biases my perspective, but mini-golf brings people together and gives them a way to connect. Everyone can play reasonably well; the competitive stakes are quite low. These qualities are common to playful experiences overall—play and tinkering puts people in a receptive frame of mind, ready to engage and take risks they might not otherwise.[8] This is true both intellectually and socially. In playful situations people are willing to try new things, pushing past hesitations they might feel in more serious settings. In playful situations people are more likely to engage with others they might not otherwise, breaking through harmful power dynamics. The playful invitation puts the participants at ease, more likely to be curious and reflective about the meaningful content. Play in these experiences isn't just a veneer on a serious data conversation, the interactions with data themselves are playful. Think back to the ladder and wall line chart, or the hole that is too small: the play is the experience of the data. In this way *The Back 9* focused centrally on playful participation as the means to engage the data and impact.

How to House invited participation by offering gallery visitors the chance to interact with the data representation itself. Like a hands-on science museum exhibit, this puts the learning in a constructive frame; people were learning by making something. This echoes pedagogy I introduced earlier, specifically related to the ideas of embodied learning and constructionism (see "How we interact with data sculptures" in Chapter 3). Visitors were taking a constructionist approach to build what Papert and Harel call a "public entity" in concert with others.[9] They collaborated to create a tangible externalized object that represented their understanding of a solution to the problem. The experiences created opportunities for connection, discussion, disagreement, and sense-making to happen out loud. The exhibit didn't give them obvious answers, but rather facilitated their exploration of the space of solutions and allowed them to build their own physical answer. This was a constructionist interactive data display built around participation.

Both experiences were designed to *build mirrors, not windows*. The LAPD hosted events at the course to invite the local public in to play. However, the word spread and soon the course was drawing other audiences from across the area. Groups such as corporate team-building retreats and city planning staff asked if they could come visit. The data about the planned zoning changes jumped off the pages of the otherwise "attractive planning document that had no power" (as artist Woo described it to me). The community were reflecting the data back to itself and others, telling a story about how the proposed zoning changes concentrated power and profit, threatening to displace the residents of Skid Row. This playable mini-golf hole was a data story designed for the very place and people represented in the data.

The *How to House* project took a complex socio-economic issue, housing, and reflected it back to the community in physical form. Earlier arguments I've made establish how low self-perception of math ability is a significant barrier to working with data in public settings; large populations of Americans just don't think they are good with numbers (see "I'm not a math person" in Chapter 3). Are you intimidated by the idea of moving lightweight boxes, stacking them on top of each other, and comparing their volumes? This is a very accessible invitation to work within mathematical constraints, computing and comparing through physical means. The exhibit functioned as a data mirror for visitors, making the math about their own community accessible and fun to engage with.

The playful design and robust data showcased in both experiences demonstrate how to *create layers of reading* when entering a complex data story. At first glance *The Back 9* looks like just a playful new experience from a place that often hosts art shows, community gatherings, and other events. People could come in and play for free without deeply engaging in the content itself. With this limited interaction, *The Back 9* simply plays a role in the larger narrative of Skid Row as a welcoming and rich community with cultural assets to enjoy. This offers a very simple first layer of reading. Of course, it'd be hard to come in and play the course without reading some of the signage on the course itself. That layer of reading introduces more of the data and history that informed the design of each hole. You'd be hard-pressed to play the chart-shaped population versus housing stock hole described previously without understanding the changes underway; you literally must climb up a ladder labeled with population growth and roll your ball down a tube showing housing stock decline. Another hole forces the player to decide which type of land to invest in, and still another explains the concept of floor area ratio (FAR), a number used in zoning laws that informs what kind of buildings get created. Simply by reading the signage you become more informed and aware of how zoning functions and what it impacts. The workshops reinforced this.

How to House similarly used recognizable forms to draw people in, augmenting explanation and introducing visitors to more data as they went through the exhibit. The initial table full of miniature people is easy to digest and hooks you in with a key number. You could leave right there and understand the main problem. However, if you continue in, the boxes each represent potential solutions to creating the needed housing, and text on the walls provides yet another layer of reading to understand the context, barriers, and potential of each solution.

The LAPD used mini-golf, dramatic performance, and participatory exhibits to *open many doors* into the data behind the threats they were facing. In *The Back 9* project the act of playing the course built experiential knowledge. Seeing the play or participating in a workshop built presentational knowledge. Reading the signs or the data visualizations included in the course built propositional knowledge. *How to House* included more invitations to build practical knowledge. The mathematical information is remixed into physical form, engaging a whole different set of audiences than numbers do. The physical media can engage those who are put off by math, or don't take to flat charts.

The Back 9 and *How to House* were designed to welcome visitors from diverse backgrounds and contexts. Activists and artists like the LAPD are leading the way in novel approaches to meeting people where they are with data stories in effective and impactful ways.

Empower communities with information

Interdisciplinary architect Buckminster Fuller opens his *Operating Manual for Spaceship Earth* by noting:

> I am enthusiastic over humanity's extraordinary and sometimes very timely ingenuity. If you are in a shipwreck and all the boats are gone, a piano top buoyant enough to keep you afloat that comes along makes a fortuitous life preserver. But this is not to say that the best way to design a life preserver is in the form of a piano top. I think that we are clinging to a great many piano tops in accepting yesterday's fortuitous contrivings as constituting the only means for solving a given problem.[10]

We're clinging to charts and graphs like piano tops in a sea of datafication and its negative impacts. Yet we aren't well served by relying on yesterday's solutions for working with and representing data when we have new sets

EMPOWER COMMUNITIES WITH INFORMATION 211

of problems in new domains. Popular data is designed to showcase the life preservers we need.

A 2023 survey of over a hundred data projects and their emotional impact offers a helpful breakdown of genres being explored: interactive interfaces, video, static images or paintings, installations, artifacts, and events.[11] The first, interactive interfaces, was the primary mechanism used in more than half of the projects, reflecting the focus on now traditional two-dimensional (2D) forms of data that I've critiqued noted this book. Beyond form, topic also matters; we've found repeatedly that people look for themselves in data and show more interest in data that is relevant to their concerns.[12]

The ideas presented so far in fact stir up parts of historical data visualization that we don't often discuss. Joseph Priestley's influential charts of history, described by some as the first modern timelines, were printed as large wall-sized prints in the 1700s.[13] The scale of them, much larger than contemporary charts from Playfair, matched Priestley's goals of stimulating critical discourse in educational settings. For hundreds of years large format visuals of data have encouraged collaborative verbal reflection. Florence Nightingale's famous charts of army mortality were part of her larger advocacy campaign to lobby for reforms.[14] She was an activist who used her influential connections to push policy responses to her data findings. The still modern-looking visualizations about the conditions of the formerly enslaved in the US presented by W. E. B Dubois at the Paris World Fair of 1900 were designed to speak with the language of power.[15] The infographics, as part of the overall exhibit, are an example of a marginalized population creating data visuals to showcase their growth and development; the team represented the cultural and economic production of the Black Americans to a global population. When we consider the contexts of some of the most popular examples of historical data visualizations, we can see that they focused on participatory, human-centered, and emancipatory goals. My focus on pro-social contexts can simply be described as returning us to some of those roots of data visualization, before it was professionalized in the sciences and statistics.

Revisiting key principles

The need for creative public impactful data experiences like the LAPD created is one I prioritize in Chapter 1, which argues that we need to *center impact* when using data in the pro-social sector. The vivid power of the Mexican data protest created by activists working with mothers of *desaparecidos*

212 CONCLUSION

shows another example of what centering impact with data looks like. I draw inspiration from the rich histories, and initial modern approaches to data, in libraries, museums, newsrooms, civil society organizations, and democratic governance. Their goals include sharing knowledge, holding power to account, building community, engaging participation, and upholding rights. These are key mechanisms for centering impact, yet our central approaches to working with data and representing it aren't optimized for these outcomes.

Chapter 2 fleshes out other critiques of our modern approaches to data science in society, which fail to *focus on participation*. The Philadelphia *How We Fish* mural offers one model of participatory data production and representation. Movements and concepts such as data for good, critical data literacy, data justice, and data feminism are alternate approaches that engage people more directly, resetting power dynamics inherent in our modern data analysis and sharing methods. I describe in detail the process of creating data murals as a case study in how to create methods where participation is central and connect it to other work showing data on walls in public. A growing global movement is using participatory data murals to put data back in the hands of the subjects it represents.

In response to the ongoing spread of datafication across society, I argue in Chapter 3 that we must *build mirrors, not windows*. The racist history of women's clothing sizes offers one example of how an earlier twentieth century rounds of social datafication created long-felt harms. The primary response from many groups to broad datafication has been to design and offer data literacy programs. These almost universally focus on technical skill building, limiting who is interested and able to engage with them, often reinforcing the idea of using data as windows. An alternative is the idea of creating data mirrors, representations that reflect data back to a community. Academic research and the artistic practice of creating data sculptures offers another novel technique for creating these mirrors. I showcase emerging approaches and practices for data sculptures that show the application in various domains, constructing concrete presentations instead of abstract ones.

The complexity of data stories turns many away, a challenge I tackle in Chapter 4 where I propose to *create layers of reading* that start simple and use various techniques to draw people into nuance and detail. The data *khangas* created by Tanzanian fashion designers are one example, building on well-known local forms of protest to invite audiences into impactful data stories while simultaneously building designer's data capacity. In contrast, I label the constraints of charts and graphs as the tyranny of the 2D, limiting our ability to imagine alternative forms of representing data outside their rules.

Research shows they don't work as consistently as we hope across diverse public audiences. An alternative approach is to use body-based understanding and somatic learning in the form of data theatre. The history of radical performance practices, put in service of data in civic contexts, is showing early promise as an effective way to engage audiences with simple invitations that can build to complex and rich understandings and dialog through data performance.

The audiences become even more central in Chapter 5, where I push to *open many doors* into data stories. The *Ripple Effect* exhibit of water contamination data introduces the idea of creating rich multi-sensory experiences to engage different types of audiences. Building on a 4-part taxonomy of ways of knowing, I delve into the fringes of non-visual data storytelling by exploring existing and potential practices of hearing, smelling, and eating your data. Sonification, smellification, and edibilization are new ways to pull people into data experiences. These can be used to reinforce other data encodings as multi-sensory experiences that open new doors for a broader set of audiences, offering more tools for telling relatable and effective data stories in community settings.

Imagine popular data in your work

Advocates fighting for community participation often use the phrase "nothing about us without us." The phrase's modern use grew from disability justice movements and has spread from there to other settings, including data activism.[16] However, it is important to remember that many use the slogan in its fuller form—"nothing about us without us is for us"—to capture the intent and goal of data collection work. Data extracted from a community ("about us") without any participation ("without us") is not truly intended to advance the community's needs ("for us"); it exists to serve someone else's agenda.

The history of data use revolves around groups in power using data to hold on to their positions and enforce control over others. This history is embedded in the processes and technologies we continue to use today. The growth of data visualization practice rests on that foundation, and in its current dominant form privileges a particular way of thinking, knowing, and representing. This especially matters in community settings, where people who don't speak that language are excluded from data-driven decision-making processes. There is another path, and it rests on practices from the world of the arts and the pro-social sector, playing with data sculptures, murals, theatre, and multi-sensory

214   CONCLUSION

experiences. We need more inclusive data storytelling, a toolkit offering potential to increase democratization, empowerment, agency, and efficacy. Driven to address the needs in the pro-social sector where data is now commonly used, I offer popular data as a new framework for working with and representing data.

This book helps you imagine and design alternatives. I find inspiration in examples such as the short epilogue of data artist Jer Thorp's *Living In Data*, where he offers a story of a future imagined community that produces and leverages data for their own needs.[17] Thorp's young fictional protagonists move across the forest collecting unauthorized environmental data, sharing it via mesh networks with their community across a near-future landscape. They even compose a sonification that merges the data with the sounds of nature to help understand its context. Thorp builds on a tradition of using fiction to imagine alternative potential futures when the forces combined against them feel overwhelming. We can dream bigger, creating examples that leverage new modes of data representation with new power, like so many I've shared throughout this book. Once you adopt this broader toolbox, the approaches and media come together in delightful, surprising, and impactful ways that let you engage with diverse audiences, creating new pathways into working with data in community settings.

I keep returning to examples from Tanzania precisely because the work there in the late 2010s was an example of imagining and creating alternatives.[18] I've already introduced their community data *khanga* initiative and fashion show (leveraging a local custom to engage broader groups around health data), their shareback sessions (bringing data back to community planning to drive engagement and efficacy), and their murals (leveraging a popular medium to meet people where they are with data messages). They went on to create a short-lived web-based data news TV show called *Data za kitaa*/Street Data, targeting youth and focused on creating "a strengthened culture of evidence-based decision making in Tanzania."[19] Episodes included animated infographics, street interviews, and host narration, all centered around using data to understand issues such as youth employment. The team also produced a song with data themed lyrics fighting against drug addiction, with a music video including data heavily sourced from the country's annual Drug Control Commission report.[20] Fashion, dialogs, murals, TV shows, music videos—this is imagining new media for data storytelling.

The groups I've highlighted here were driven to explore alternatives because decisions that involve community input and impact in public settings are increasingly centered around data in some way. The current norms limit

EMPOWER COMMUNITIES WITH INFORMATION 215

our options when we pivot from business and science settings into collective decision-making processes in local or global pro-social contexts. Participatory data practices are emerging across the globe. Scientific ways of knowing are being shown to be just one of many ways of building knowledge. Data is being taken to the streets to meet people where they are. Our understanding of how people interact with data to take away simple messages, as well as complex nuanced narratives, is growing. People are asking "so what?" when they look at a chart, and critically considering the impact of data insights.

US Congresswoman Ayanna Pressley eloquently argues that "the people closest to the pain should be the closest to the power."[21] Those that are most likely to be impacted by data-connected processes should be in positions of power to make decisions. Popular data is a framework to help make that happen in datafied community settings. We can bring people together around data, driven by a moral imperative to offer more people a seat at the table in decision making processes that impact them. It's up to us to push the boundaries of who and what data is for. Like the title of one of my early workshops, together we must "Fight the Bar Chart." The popular data framework offers a larger, more creative, toolkit. Add this to your tool belt and you'll find that it helps you empower those you work with and for, building a more just datafied world. Together, we can do better. The diverse examples and passionate practitioners cataloged throughout this book leave me with hope that we can retake some of the power we've embedded in data and build more justice-centric approaches to using it.

Acknowledgments

This book is the individual output of a community of like-minded thinkers. I am so grateful to the many people, communities, and organizations that have inspired and collaborated with me, co-creating the approaches, ideas, and projects I've shared throughout this book.

My wife Emily not only collaborated and inspired so many of these projects, but also launched me down this path of inquiry and work so many years ago by simply asking if I could run a workshop for the communities she worked with. Her generous support, creativity, intelligence, and love amplify and underlie all my work. My two kids have contributed to most of these projects as well, from collecting spoons, to mixing paint, to asking challenging questions about my silly ideas; I owe so much to both of you for your prompts and conversation and love you so much. I come from a long line of authors who have created a path to producing knowledge in this way for me to learn from and continue; I owe a great debt to my grandfather, father, mother, and brother.

Colleagues from my time at the Massachusetts Institute of Technology (MIT) Center for Civic Media have been inspirational partners on this path of research and impact. To Ethan Zuckerman I owe thanks for the space to pursue this work under the umbrella of an official "MIT Project," creating significant opportunities for impact, networking, and validation. To Catherine D'Ignazio I owe so much for coming in as a collaborator, conspirator, and colleague. Our ideas and approaches are intertwined at various points, and I've learned so much from her theories and reflections. To Laura Perovich and Vicky Palacin Silva I offer my thanks for launching my inquiry into embodiment and data; your questions and provocations led me to new approaches. To the large group that gathered around the weekly C4CM meeting table I offer my thanks for humoring my ideas (good and bad) by lending me your time and brains to try out activities—thanks for your encouragement and your critique.

These projects are collaborations with many organizations, each lending me their time, expertise, wisdom, and passion for their causes. Doctors for Global Health let us try out our first data mural at their annual convening. The Boy and Girls Club of Cambridge, Groundwork Somerville, Lisa Brukilacchio, Somerville Head Start, the Government of Minas Gerais, Plug Minas, the Curry School in Boston, and the Somerville Food Security Coalition have all supported the Data Murals work. Numerous organizations have helped me refine my ideas and practice, including the Regional Center for Healthy Communities, Guiding Lights, Hacks/Hackers, Pen Plus Bytes, the Boston Foundation, NetSquared, Boston Area Research Initiative (BARI), New Sector Alliance, Allied Media, Harvard Meta Lab, Open Knowledge Foundation, Harvard Systemic Justice Project, MIT Open-DocLab, Harvard Graduate School of Design (GSD), Central Massachusetts Regional Planning Council, Stanford Center on Philanthropy and Civil Society, Boston University, Bloomberg, TechNetworks Boston, YouthArts Boston, MIT Museum, DataPop Alliance, Third Sector New England, New York University (NYU), United Nations (UN) World Data Forum, Draper Richards Kaplan (DRK) Foundation, Massachusetts Technical Assistance Partnership for Prevention, the Workshop school in Merida, World Food Program, Dokk1, Data for Black Lives, The Guardian, Boston Public Libraries, Saint Paul Public

ACKNOWLEDGMENTS 217

Library, Qlik, National Academies of Science, Engineering, and Medicine, Massachusetts Community and Banking Council, and the National Library of Medicine.

Like all research, this work costs money to pull off. Besides the in-kind support of those collaborating groups, I owe thanks to supporters at the Knight Foundation, Making all Voices Count, MIT and Northeastern for the financial contributions.

This work engages with many other domains and disciplines, so I'd be remiss to not thank the artists, historians, designers, activists, and technologists whose work is mentioned or described throughout. A critical goal of mine is to create a more culturally, geographically, and topically diverse set of inspirational examples for those working with data, and it has been a delight to be personally inspired by all the new examples and ideas I've found in your work. For this book I interviewed Rosten Woo, Henriëtte Brouwers, John Malpede, and Martha Muñoz Aristizabal, Dorsey Kaufmann, and Ennis Carter, all of whom shared their wisdom, experiences, and media generously. Of course, finding community around academic research at the intersection of fields can be challenging. I'd like to thank the gathering spaces who have welcoming my work—Institute of Electrical and Electronics Engineers (IEEE) Vis, Association of Internet Researchers (AoIR), OpenVis Conf, University of Hyderabad, Gordon Research Conferences (GRC) Visualization, Designing Interactive Systems (DIS), the American Public Health Association (APHA), WebSci, and the Impacts of Civic Technology Conference (TICTeC).

Over the last few years at Northeastern I've found a new set of colleagues who I can thank foremost for not laughing me out of the room. I entered as a wacky guy using data to paint murals, devise theatre, and build sculptures, and each of them seems to have nodded to themselves and said "yeah, that makes sense," which is not the response I expected but is one I've delighted in. To Jesse Hinson and Amanda Brea I offer my thanks for jumping in with this unknown person to explore early ideas of data theatre, and for sharing your deep love and knowledge of the discipline I was stepping into with naïveté. To Dani Snyder-Young, Michael Arnold Mages, Jonathan Carr, Oliver Wason, Moira Zellner, Victor Talmadge, and George Belliveau I offer the thanks for your energy and patience with me as I delve further into the field of theatre and learn more from your wisdom. Your collaboration has been vital, and I thank you for it. More broadly my Northeastern colleagues in Journalism and Art + Design have been an amazing group to lean on and helped push this work forward.

A large group of people have helped review drafts of various sections of this book. I owe thanks and appreciation for generous time and thoughtful comments to Ethan Zuckerman, Martha Muñoz Aristizabal, Dorsey Kaufmann, Ennis Carter, Henriëtte Brouwers, John Malpede, Jane Kunze, Laura Perovich, Sameul Huron, Enrico Bertini, Jonathan Carr, Camilia Matuk, Alberto Cairo, Catherine D'Ignazio (and her Data + Feminism Lab), Emily Bhargava, and Allen Downey. A special thanks to Benjamin Zi-Hao for his research assistance on non-visual data storytelling, and to Nicole Wilder for support on permissions acquisitions.

I hope these acknowledgements demonstrate how this book is the output of a dispersed community of practice, built on the energy, excitement, and enthusiasm of those groups and more. I'm so grateful for the generosity of so many and hope to continue with new collaborators who see resonant ideas and practices within this text.

References

Introduction

1. This title, "Got Data?," was a reference to the long-running "Got Milk?" advertising campaign created by the US Dairy industry.
2. Resnick, M. (2018). *Lifelong Kindergarten: Cultivating Creativity through Projects, Passion, Peers, and Play.* The MIT Press. Cambridge, MA.
3. D'Ignazio, C. & R. Bhargava. (2016). "DataBasic: Design Principles, Tools and Activities for Data Literacy Learners." *The Journal of Community Informatics*, 12(3), 83–107.
4. Noble, S. U. (2018). *Algorithms of Oppression: How Search Engines Reinforce Racism.* (Illustrated edition). NYU Press. New York, NY.
5. O'Neil, C. (2016). *Weapons of Math Destruction: How Big Data Increases Inequality and Threatens Democracy.* Crown. New York, NY.
6. Eubanks, V. (2018). *Automating Inequality: How High-Tech Tools Profile, Police, and Punish the Poor.* St. Martin's Press. New York, NY.
7. Benjamin, R. (2019). *Race After Technology: Abolitionist Tools for the New Jim Code.* Polity. Cambridge, UK.
8. Perez, C. C. (2021). *Invisible Women: Data Bias in a World Designed for Men.* Abrams. New York, NY.
9. D'Ignazio, C. & L. F. Klein. (2020). *Data Feminism.* The MIT Press. Cambridge, MA.
10. Hall, P. A. & P. Dávila. (2022). *Critical Visualization: Rethinking the Representation of Data.* Bloomsbury. New York, NY.
11. Williams, S. (2020). *Data Action: Using Data for Public Good.* The MIT Press. Cambridge, MA.
12. Loukissas, Y. A. (2022). *All Data Are Local.* The MIT Press.Cambridge, MA. https://mitpress.mit.edu/9780262545174/all-data-are-local/.
13. H.R.1770–115th Congress (2017–2018): OPEN Government Data Act. (2019). http://www.congress.gov/bill/115th-congress/house-bill/1770.
14. Gitelman and Jackson similarly describe the subjective nature of data production as akin to how a photographer frames the subject of their photo. Gitelman, L. & V. Jackson. (2013). "Introduction." In *"Raw Data" Is an Oxymoron*, L. Gitelman (ed.), 5. The MIT Press. Cambridge, MA. https://mitpress.mit.edu/9780262518284/raw-data-is-an-oxymoron/.
15. Popularized in Morozov, E. (2014). *To Save Everything, Click Here* (reprint edn). PublicAffairs. New York, NY.
16. Childs, M. (2023, January 4). "What Happens Now to Effective Altruism, SBF's Pet Philosophy?" *Town & Country*. https://www.townandcountrymag.com/society/money-and-power/a42330166/effective-altruism-sam-bankman-fried-ftx-fallout/.
17. Galeano, E. (1992). *The Book of Embraces* (C. Belfrage, Trans.; reprint edn). W. W. Norton & Company. New York, NY.

Chapter 1

1. These messages from mothers are selected from a spreadsheet of all the responses on the WhatsApp group that Muñoz Aristizabal shared with me.
2. Morland, S. (2023, May 12). "Mexico to Launch Database of over 100,000 'Disappeared' People." *Reuters*, https://www.reuters.com/world/americas/mexico-launch-database-over-100000-disappeared-people-2023-05-12/.
3. D'Ignazio, C. (2022). *Counting Feminicide: Data Feminism in Action.* MIT Press. Cambridge, MA. https://mitpressonpubpub.mitpress.mit.edu/counting-feminicide, chapter 6.
4. Ricaurte, P. (2019). "Data Epistemologies, The Coloniality of Power, and Resistance." *Television & New Media*, 20(4), 350–365. https://doi.org/10.1177/1527476419831640.
5. Roosen, L. J., C. A. Klöckner & J. K. Swim. (2018). Visual Art as a Way To Communicate Climate Change: A Psychological Perspective on Climate Change-Related Art." *World Art*, 8(1), 85–110, https://doi.org/10.1080/21500894.2017.1375002.

REFERENCES 219

6. Bowler, L., A. Acker, & Y. Chi. (2019). "Perspectives on Youth Data Literacy at the Public Library: Teen Services Staff Speak Out." *Journal of Research on Libraries & Young Adults*, 10(2).
7. Devlin, H. (2023, January 5). "Amateur Archaeologist Uncovers Ice Age 'Writing' System." *The Guardian*, https://www.theguardian.com/science/2023/jan/05/amateur-archaeologist-uncovers-ice-age-writing-system.
8. Bacon, Bennett, Azadeh Khatiri, James Palmer, Tony Freeth, Paul Pettitt, and Robert Kentridge. (2023). "An Upper Palaeolithic Proto-Writing System and Phenological Calendar." *Cambridge Archaeological Journal*, January 1–19, https://doi.org/10.1017/S0959774322000415.
9. Marchese, F. T. (2011). "Exploring the Origins of Tables for Information Visualization." 2011 15th International Conference on Information Visualisation, 395–402, https://doi.org/10.1109/IV.2011.36.
10. Robson, E. (2003). "Tables and Tabular Formatting in Sumer, Babylonia, and Assyria, 2500 BCE–50 CE." In *The History of Mathematical Tables: From Sumer to Spreadsheets*, M. Campbell-Kelly, M. Croarken, R. Flood, & E. Robson (eds.), 19–47, at 24. Oxford University Press Cambridge, UK, https://doi.org/10.1093/acprof:oso/9780198508410.001.0001.
11. Campbell-Kelly, M., M. Croarken, R. Flood, & E. Robson. (eds.). (2003). *The History of Mathematical Tables: From Sumer to Spreadsheets*. Oxford University Press, Cambridge, UK. https://doi.org/10.1093/acprof:oso/9780198508410.001.0001.
12. Rosenberg, Daniel. (2018). "Data as Word." *Historical Studies in the Natural Sciences*, 48(5), 557–567, https://doi.org/10.1525/hsns.2018.48.5.557; Rosenberg, Daniel. (2021). "Data." In *Information—a Historical Companion*, Ann Blair, Duguid, Anja-Silvia Goeing, and Anthony Grafton, (eds.), 387–391. Princeton University Press. Princeton, NJ.
13. Porter, Theodore M. (2020). *The Rise of Statistical Thinking, 1820–1900*. Princeton University Press. Princeton, NJ.
14. Sutherland, I. (2005). "Graunt, John." In *Encyclopedia of Biostatistics*. John Wiley & Sons, Ltd. Hoboken, NJ, https://doi.org/10.1002/0470011815.b2a17055.
15. Donoho, D. (2017). "50 Years of Data Science." *Journal of Computational and Graphical Statistics*, 26(4), 745–766, https://doi.org/10.1080/10618600.2017.1384734; Cao, L. (2017). "Data Science: A Comprehensive Overview." *ACM Computing Surveys*, 50(3), 1–42, https://doi.org/10.1145/3076253.
16. Cleveland, William S. (2001). "Data Science: An Action Plan for Expanding the Technical Areas of the Field of Statistics." *International Statistical Review*, 69(1), 21–26, https://doi.org/10.1111/j.1751-5823.2001.tb00477.x.
17. Boyd, Danah & K. Crawford. (2012). "Critical Questions for Big Data: Provocations for a Cultural, Technological, and Scholarly Phenomenon." *Information, Communication & Society*, 15(5), 662–679, https://doi.org/10.1080/1369118X.2012.678878.
18. D'Ignazio, C. & R. Bhargava. (2015, September 28). *Approaches to Building Big Data Literacy*. Bloomberg Data for Good Exchange. New York, NY, USA: Bloomberg Data for Good Exchange.
19. "Look Inside the Most Cutting-Edge Public Library in the World." (2016, August 19). *Time*, https://time.com/4458185/denmark-library-cutting-edge-dokk1/.
20. Fagan, L. (2019, December 31). "In Aarhus, a Bell Tolls for New Babies—and Calls us To Action." *Sustainability Times*, https://www.sustainability-times.com/clean-cities/in-aarhus-a-bell-tolls-for-new-babies-and-calls-us-to-action/.
21. Kunze, J. (2020). "Data Literacy in the Smart City." *Geoforum Perspektiv*, 19(35), article 35, https://doi.org/10.5278/ojs.perspektiv.v19i35.6423.
22. Visit https://www.nextlibrary.net to learn more about their programs.
23. Patin, B. (2020). "What Is Essential?: Understanding Community Resilience and Public Libraries in the United States During Disasters." *Proceedings of the Association for Information Science and Technology*. Association for Information Science and Technology, 57(1), e269, https://doi.org/10.1002/pra2.269.
24. Boston Public Library. (2011, November 15). *Strategic Plan*. https://www.bpl.org/about-the-bpl/strategic-plan/.
25. Khan, H. R. & Y. Du. (2017). *What Is a Data Librarian?: A Content Analysis of Job Advertisements for Data Librarians in the United States Academic Libraries*, https://library.ifla.org/id/eprint/2255/; Ohaji, I. K., B. Chawner, & P. Yoong. (2019, December 15). *The Role of a Data Librarian in Academic and Research Libraries.*, Information Research, 24(4), paper 844. Retrieved from http://InformationR.net/ir/24-4/paper844.html;Rice, R. & J. Southall. (2016). *The Data Librarian's Handbook*. Facet Publishing, London, UK. https://www.alastore.ala.org/content/data-librarians-handbook-hardcover.

220 REFERENCES

26. King, G. (2007). "An Introduction to the Dataverse Network as an Infrastructure for Data Sharing." *Sociological Methods & Research*, 36(2), 173–199, https://doi.org/10.1177/0049124107306660.

27. Abilock, Debbie, Susan D. Ballard, Tasha Bergson-Michelson, Jennifer Colby, Catherine D'Ignazio, Kristin Fontichiaro, et al. (2017). *Data Literacy in the Real World: Conversations & Case Studies*, A. L. Kristin Fontichiaro (ed.). Ann Arbor, MI: Michigan Publishing, University of Michigan Library, http://dx.doi.org/10.3998/mpub.9970368.

28. Bowler, L., A. Acker, & Y. Chi. (2019). "Perspectives on Youth Data Literacy at the Public Library: Teen Services Staff Speak out." *Journal of Research on Libraries and Young Adults*, 10(2), 1–21.

29. Civic Switchboard Project. (2019). "Connecting Libraries and Civic Data," https://civic-switchboard.github.io/assets/guide/Connecting_Libraries_and_Civic_Data.pdf.

30. Greaves, W. (Director). (1989, December 19). Ida B. Wells. "A Passion for Justice." In *American Experience.*

31. Lauderdale, V. (2021, June 9). "From the Vault: Building the Frisco Bridge." *Memphis Magazine.* https://memphismagazine.com/api/content/3108dbf6-bb64-11eb-9991-1244d5f7c7c6/.

32. Rose, M. A. (2020, January 27). *St. Patrick's Day Snowstorm of 1892.* National Weather Service; NOAA's National Weather Service. https://www.weather.gov/ohx/18920317.

33. Tucker, D. M. (1971). "Miss Ida B. Wells and Memphis Lynching." *Phylon—the Atlanta University Review of Race and Culture*, 32(2), 112–122, https://doi.org/10.2307/273997.

34. Simpson, M. (2004). "Archiving Hate: Lynching Postcards at the Limit of Social Circulation." *English Studies in Canada*, 30(1), 17–38.

35. *Ida B Wells s A Red Record | The New York Public Library.* (n.d.). Retrieved June 8, 2023, from https://www.nypl.org/events/exhibitions/galleries/beginnings/item/3548.

36. McCombs, M. E. & D. L. Shaw. (1972). "The Agenda-Setting Function of Mass Media." *The Public Opinion Quarterly*, 36(2), 176–187, https://doi.org/10.1086/267990.

37. Bateson, G. (2000). *Steps to an Ecology of Mind: Collected Essays in Anthropology, Psychiatry, Evolution, and Epistemology.* University of Chicago Press. Chicago, IL. https://press.uchicago.edu/ucp/books/book/chicago/S/bo3620295.html.

38. Schudson, M. (2008). *Why Democracies Need an Unlovable Press.* Polity. Cambridge, UK.

39. Fioroni, S. (2022, May 19). "Local News Most Trusted in Keeping Americans Informed About Their Communities." *Knight Foundation—Learning and Impact.* https://knightfoundation.org/articles/local-news-most-trusted-in-keeping-americans-informed-about-their-communities/; Kennedy, D. (2021, February 23). "'Mogul Roulette,' Or The Totally Random Destruction Of Local News." *GBH.* https://www.wgbh.org/news/commentary/2021-02-23/mogul-roulette-or-the-totally-random-destruction-of-local-news.

40. Kantor, A. & S. Rafaeli. (2021). "Independence Through Data Journalism." AoIR Selected Papers of Internet Research, https://doi.org/2013; Segel, E. &, J. Heer. (2010). "Narrative Visualization: Telling Stories with Data." *Visualization and Computer Graphics, IEEE Transactions On*, 16(6), 1139–1148.

41. Green, E., Holliday, D., & M. Rispoli. (2023). "The Roadmap for Local News: An Emergent Approach to Civic Information Needs," https://localnewsroadmap.org/.

42. Dick, M. (2020). *The Infographic.* MIT Press. Cambridge, MA.

43. Müller, N. C. & J. Wiik. (2023). "From Gatekeeper to Gate-Opener: Open-Source Spaces in Investigative Journalism." *Journalism Practice*, 17(2), 189–208, https://doi.org/10.1080/17512786.2021.1919543.

44. Tracy, M. (2019, December 1). "These Reporters Rely on Public Data, Rather Than Secret Sources." *The New York Times*, https://www.nytimes.com/2019/12/01/business/media/open-source-journalism-bellingcat.html.

45. The Washington Post. (n.d.). "Fatal Force." *Washington Post.* Retrieved September 5, 2023, from https://www.washingtonpost.com/graphics/investigations/police-shootings-database/.

46. Mattu, J. A. and Surya Lauren Kirchner. (n.d.). "Minority Neighborhoods Pay Higher Car Insurance Premiums Than White Areas With the Same Risk." *ProPublica.* Retrieved September 5, 2023, from https://www.propublica.org/article/minority-neighborhoods-higher-car-insurance-premiums-white-areas-same-risk.

47. Lowery, W. (2020, June 23). "A Reckoning Over Objectivity, Led by Black Journalists." *The New York Times*, https://www.nytimes.com/2020/06/23/opinion/objectivity-black-journalists-coronavirus.html.

48. Crandon, G. (1990). "Media View of the Police." *Policing*, 6(3), 573–581.

REFERENCES   221

49. Tong, J. (2022). "The Rise of Partisan Journalism and the Crisis of Objective Journalism." In *Journalism, Economic Uncertainty and Political Irregularity in the Digital and Data Era*, 93–107, Emerald Publishing Limited. Bingley, https://doi.org/10.1108/978-1-80043-558-220221007.

50. Folkenflik, D. (2020, June 8). "Editors Barred a Black Reporter from Covering Protests. Then her Newsroom Rebelled ." *NPR*, https://www.npr.org/2020/06/08/872234014/editors-barred-a-black-reporter-from-covering-protests-then-her-newsroom-rebelle.

51. Moser, S. C. (2010). "Communicating Climate Change: History, Challenges, Process and Future Directions." *Wiley Interdisciplinary Reviews: Climate Change*, 1(1), 31–53.

52. Sterman, John et al. (2012). "Climate Interactive: The C-ROADS Climate Policy Model." *System Dynamics Review*, 28(3), 295–305.

53. Allen, S. (2004). "Designs for Learning: Studying Science Museum Exhibits that Do more than Entertain." *Science Education*, 88(S1), S17–S33, https://doi.org/10.1002/sce.20016.

54. Melhuish, F. (2020, May 12). *A Cabinet of Curiosities: Ole Worm's 'Museum Wormianum' (1655)*. University of Reading—Special Collections, https://collections.reading.ac.uk/special-collections/2020/05/12/a-cabinet-of-curiosities-ole-worms-museum-wormianum-1655/.

55. Prottas, N. (2019). "Where Does the History of Museum Education Begin?" *Journal of Museum Education*, 44(4), 337–341, https://doi.org/10.1080/10598650.2019.1677020.

56. Tate, N. B. (2012). "Museums as Third Places or What? Accessing the Social Without Reservations." *Museums & Social Issues*, 7(2), 269–283, https://doi.org/10.1179/msi.2012.7.2.269.

57. Dana, J. C. (1999). *The New Museum: Selected Writings by John Cotton Dana* (W. Peniston, Ed.). American Association of Museums. Washington, D.C.

58. Weil, S. E. (2002). *Making Museums Matter* (first edn). Smithsonian Books. Washington, DC.

59. Cole, K. C. & M. Gell-Man. (2012). *Something Incredibly Wonderful Happens: Frank Oppenheimer and His Astonishing Exploratorium* (reprint edn). University of Chicago Press. Chicago, IL.

60. Gurian, E. H. (2005). "Function Follows Form: How Mixed-Used Spaces in Museums Build Community." In *Civilizing the Museum*. Routledge. London, UK.

61. Simon, N. (2010). *The Participatory Museum*. Museum 2.0 (first edition), https://participatorymuseum.org/.

62. Disclosure: I served on the board of Tactical Tech Collective when the Glass Room work was starting but was not involved in its creation.

63. Leonard, R. (1999). "'Seeing Is Believing': Otto Neurath, Graphic Art, and the Social Order." *History of Political Economy*, 31, 452–478.

64. Badenoch, A. (n.d.). "A Museum without Borders—Inventing Europe." *Inventing Europe*. Retrieved November 27, 2023, from https://www.inventingeurope.eu/story/a-museum-without-borders.

65. I appreciate the Neurath's contributions while still critiquing the strong racist stereotypes that permeated their visual depictions of silhouetted people from around the world (all their Indians wear a turban).

66. Baack, Stefan. (2015). "Datafication and Empowerment: How the Open Data Movement Re-Articulates Notions of Democracy, Participation, and Journalism." *Big Data & Society* 2 (2), 2053951715594634, https://doi.org/10.1177/2053951715594634.

67. Malamud, C. (2007, October 22). "Open Government Working Group: Memorandum." *Public.Resource.Org*, https://public.resource.org/open_government_meeting.html.

68. Porter, T. (2020). *Trust in Numbers*. Princeton University Press, Princeton, NJ. https://press.princeton.edu/books/paperback/9780691208411/trust-in-numbers.

69. Hintz, A., D. Dencik, J. Redden, E. Treré, J. Brand, & H. Warne (2022). *Civic Participation in the Datafied Society: Towards Democratic Auditing?* Data Justice Lab. Cardiff, UK. https://datajusticelab.org/wp-content/uploads/2022/08/CivicParticipation_DataJusticeLab_Report2022.pdf.

70. Milner, Y. (2019, March 27). "Abolish Big Data. United Nations 24th Working Group for the International Decade for People of African Descent." Geneva, Switzerland, https://medium.com/@YESHICAN/abolish-big-data-ad0871579a41.

71. Heyneman, S. P., J. P. Farrell, & M. A. Sepulveda-Stuardo. (1978). *Textbooks and Achievement: What we Know* (World Bank Staff Working Paper No. 298). World Bank Washington, DC.

72. UNESCO. (2013). *Textbooks and Learning Resources: A Global Framework for Policy Development* (Education Sector). United Nations Educational, Scientific and Cultural Organization. Paris, FR. https://unesdoc.unesco.org/ark:/48223/pf0000232222.

222 REFERENCES

73. Fox, J. (2001). "Vertically Integrated Policy Monitoring: A Tool for Civil Society Policy Advocacy." *Nonprofit and Voluntary Sector Quarterly—NONPROFIT VOLUNT SECT Q*, 30, 616–627, https://doi.org/10.1177/0899764001303015.
74. Graeff, E. (2019). "Monitorial Citizenship." In *The International Encyclopedia of Media Literacy*, 1–15. John Wiley & Sons, Ltd Hoboken, NJ. https://doi.org/10.1002/9781118978238.ieml0169; Schudson, M. (2011). *The Good Citizen: A History of American Civic Life* (reprint edn). Free Press. New York, NY. 310.
75. Historical context for this term can be traced from German philosopher George Hegel to French aristocrat Alexis de Tocqueville. This Euro-centric formulation has found broader application over time, especially once institutionalized via formal global bodies.
76. NGOs and CSOs: A Note on Terminology. (2013). In *Working with Civil Society in Foreign Aid*. United Nations Development Programme. New York, NY. https://www.undp.org/sites/g/files/zskgke326/files/migration/cn/UNDP-CH03-Annexes.pdf.
77. Nucera, D. (ed.). (2015). "Opening Data (p. ')." Detroit Community Technology Project, http://detroitcommunitytech.org/sites/default/files/librarypdfs/opening_data.pdf; Nucera, D. & K. Sonnenberg (eds.). (n.d.). Opening Data 2. Detroit Community Technology Project. Retrieved November 14, 2023, from https://detroitcommunitytech.org/system/tdf/librarypdfs/OpeningData-Vol2.pdf.
78. Two examples of books featuring Algoe are John Green's 2008 novel *Paper Towns* (later made into a movie) and Peng Shepard's 2022 book *The Cartographers*.
79. Chapin, M., Z. Lamb, & B. Threlkeld. (2005). "Mapping Indigenous Lands." *Annual Review of Anthropology*, 34(1), 619–638, https://doi.org/10.1146/annurev.anthro.34.081804.120429.

Chapter 2

1. Rochfort, D. (1998). *Mexican Muralists: Orozco, Rivera, Siqueiros*. San Francisco, California: Chronicle Books.
2. Cockcroft, E. S. & H. Barnet-Sánchez (eds.). (1993). *Signs from the Heart: California Chicano Murals*, (pbk. edn). Social and Public Art Resource Center. Albuquerque, New Mexico: University of New Mexico Press.
3. Golden, J., R. Rice, M. Y. Kinney, D. Graham, & J. Ramsdale. (2002). *Philadelphia Murals & Stories They Tell* (first edition). Philadelphia, PA: Temple University Press.
4. Stern, M. J. & S. C. Seifert. (2003). *An Assessment of Community Impact of the Philadelphia Department of Recreation Mural Arts Program* (Social Impact of the Arts Project). Philadelphia, PA: University of Pennsylvania School of Social Work, https://repository.upenn.edu/siap_mural_arts/, 67.
5. Mural Arts Philadelphia (Director). (2012, September 13). "Newsworks: How We Fish Labor Day Event," https://www.youtube.com/watch?v=flujdwp8Qvg.
6. Setiawan, T. (2010). *Role of Public Art in Urban Environment: A Case Study of Mural Art in Yogyakarta City*. Yogyakarta, Indonesia: Universitas Gadjah Mada, http://etd.repository.ugm.ac.id/home/detail_pencarian/48386.
7. Adams, J. (2002). "Art in Social Movements: Shantytown Women's Protest in Pinochet's Chile." *Sociological Forum*, 17, 21–56. http://link.springer.com/article/10.1023/A:1014589422758.
8. Tebes, J. L., S. L. Matlin, B. Hunter, A. B. Thompson, D. M. Prince, & N. Nohatt. (2015). *Porch Light Program: Final Evaluation Report*. New Haven, Connecticut: Yale University School of Medicine.
9. Petronienė, S. & S. Juzelėnienė. (2022). "Community Engagement via Mural Art to Foster a Sustainable Urban Environment." *Sustainability*, 14(16), article 16, https://doi.org/10.3390/su141610063.
10. Lee, B., N. Henry Riche, P. Isenberg, & S. Carpendale. (2015). "More than Telling a Story: A Closer Look at the Process of Transforming Data into Visually Shared Stories." *IEEE Computer Graphics and Applications*, 35(5), 84–90.
11. Bradshaw, P. (2011, July 7). "The Inverted Pyramid of Data Journalism." *Online Journalism Blog*, http://onlinejournalismblog.com/2011/07/07/the-inverted-pyramid-of-data-journalism/.
12. Gabrys, J., H. Pritchard, & B. Barratt. (2016). "Just Good Enough Data: Figuring Data Citizenships through Air Pollution Sensing and Data Stories." *Big Data & Society*, 3(2), 2053951716679677, https://doi.org/10.1177/2053951716679677.
13. Aisch, G., A. Cox, & K. Quealy. (2015, May 24). "You Draw It: How Family Income Predicts Children's College Chances." *The New York Times*, https://www.nytimes.com/interactive/2015/05/28/upshot/you-draw-it-how-family-income-affects-childrens-college-chances.html.

REFERENCES 223

14. Buchanan, L., H. Park, & A. Pearce. (2017, January 15). "You Draw It: What Got Better or worse During Obama's Presidency." *The New York Times*, https://www.nytimes.com/interactive/2017/01/15/us/politics/you-draw-obama-legacy.html.

15. Katz, J. (2017, April 14). "You Draw It: Just how Bad Is the Drug Overdose Epidemic?" *The New York Times*, https://www.nytimes.com/interactive/2017/04/14/upshot/drug-overdose-epidemic-you-draw-it.html.

16. Agbali, M., C. Trillo, Y. Arayici, & T. Fernandod. (2017). "Creating Smart and Healthy Cities by Exploring the Potentials of Emerging Technologies and Social Innovation for Urban Efficiency: Lessons from the Innovative City of Boston." *International Journal of Civil, Environmental, Structural, Construction and Architectural Engineering*, 11(5), 600–610, https://salford-repository.worktribe.com/preview/1490268/Smart%20and%20Healthy%20City%20Case%20Study%20Boston.pdf.

17. McDowell, C. (2015, June 11). "Big Democracy: All In." *Interaction Institute for Social Change*, https://interactioninstitute.org/all-in/.

18. Adinani, H. (2019, January 2). "We Borrowed Data from Citizens. Now, we're Giving it Back." *Data Zetu*, https://medium.com/data-zetu/we-borrowed-data-from-citizens-now-were-giving-it-back-c1d256cdbc52.

19. Couldry, N. & U. A. Mejias. (2019). "Data Colonialism: Rethinking Big Data's Relation to the Contemporary Subject." *Television & New Media*, 20(4), 336–349, https://doi.org/10.1177/1527476418796632.

20. Murray, B., E. Falkenburger, & P. Saxena. (2015). *Data Walks: An Innovative Way to Share Data with Communities*. Washington, DC: Urban Institute, https://policycommons.net/artifacts/632210/data-walks/1613522/.

21. Washington, A. A. L. (2023). *Ethical Data Science: Prediction in the Public Interest*. Oxford, UK: Oxford University Press.

22. Mastercard. (2020, January 22). "Data.org A Platform for Data Science Partnerships." *The Center for Inclusive Growth*, https://www.mastercardcenter.org/press-releases/dataorg-a-platform-for-data-science-partnerships.

23. National Academies of Sciences, Engineering, and Medicine. (2020). "Meeting #9: Motivating Data Science Education through Social Good." In *Roundtable on Data Science Postsecondary Education: A Compilation of Meeting Highlights*, 223. Washington, DC: National Academics Press, https://doi.org/10.17226/25804.

24. Hooker, S. (2018, July 22). "Why 'Data for Good' Lacks Precision." *Towards Data Science*, https://towardsdatascience.com/why-data-for-good-lacks-precision-87fb48e341f1.

25. Muyoya, C., A. Jimenez Cisneros, & R. Železný-Green. (2022, May 24). "6 Steps to Get Started on Decolonizing Data for Development." *Data.Org*, https://data.org/news/decolonizing-data-for-development/.

26. Viera Magalhães, J. & N. Couldry. (2021). "Giving by Taking Away: Big Tech, Data Colonialism and the Reconfiguration of Social Good." *International Journal of Communication*, 15, 343–362.

27. Ricaurte, P. (2019). "Data Epistemologies, The Coloniality of Power, and Resistance." *Television & New Media*, 20(4), 350–365, https://doi.org/10.1177/1527476419831640.

28. Davies, T. (n.d.). "Connected by Data | Participatory Data Governance in Practice." Connected By Data. Retrieved November 14, 2023 from https://connectedbydata.org/cases.

29. Carroll, S. C., I. Garba, O. L. Figueroa-Rodríguez, J. Holbrook, R. Lovett, S. Materechera, et al. 2020. "The CARE Principles for Indigenous Data Governance." *Data Science Journal*, 19(43), 1–12, https://doi.org/10.5334/dsj-2020-043.

30. Markham, A. N. (2019). "Critical Pedagogy as a Response to Datafication." *Qualitative Inquiry*, 25(8), 754–760, https://doi.org/10.1177/1077800418809470.

31. Deahl, E. S. (2014). "Better the Data You Know: Developing Youth Data Literacy in Schools and Informal Learning Environments." Cambridge, Massachusetts: Massachusetts Institute of Technology, http://papers.ssrn.com/sol3/papers.cfm?abstract_id=2445621.

32. Lim, V., E. Deahl, L. Rubel, & S. Williams. (2015). "Local Lotto: Mathematics and Mobile Technology to Study the Lottery." In *Cases on Technology Integration in Mathematics Education*, 43–67. Pennsylvania: IGI Global, https://doi.org/10.4018/978-1-4666-6497-5.ch003.

33. Bowler, L., A. Acker, & Y. Chi. (2019). "Perspectives on Youth Data Literacy at the Public Library: Teen Services Staff Speak Out." *The Journal of Research on Libraries and Young Adults*, 10(2).

34. Louie, J. (2022). *Critical Data Literacy: Creating a More Just World with Data*. (Workshop on Foundations of Data Science for Students in Grades K–12). Washington, DC: National Academy of Sciences.

224 REFERENCES

35. Milner, Y. & A. Traub. (2021). Data Capitalism and Algorithmic Racism, 43. https://www.demos.org/research/data-capitalism-and-algorithmic-racism.
36. Mattu, S., J. Angwin, L. Kirchner, & J. Larson. (2016). "Machine Bias." *ProPublica*. New York: Demos, https://www.propublica.org/article/machine-bias-risk-assessments-in-criminal-sentencing.
37. Nucera, D. & K. Sonnenberg (eds.). (n.d.). "Opening Data 2. Detroit Community Technology Project." Detroit, MI. Retrieved November 14, 2023, from https://detroitcommunitytech.org/system/tdf/librarypdfs/OpeningData-Vol2.pdf.
38. Nucera, D. (ed.). (2015). "Opening Data (p. ')." Detroit Community Technology Project. Detroit, MI. http://detroitcommunitytech.org/sites/default/files/librarypdfs/opening_data.pdf (Page 43).
39. Fraser, N. (2008). "Abnormal Justice." *Critical Inquiry*, 34(3), 393–422, at 405. https://doi.org/10.1086/589478.
40. re:publica (Director). (2017, May 11). *Re:publica 2017—Towards Data Justice: Social Justice in the Era of Datafication*. https://www.youtube.com/watch?v=AGqGft1k2us.
41. Independent Expert Advisory Group on a Data Revolution for Sustainable Development. (2014). *A World that Counts*. New York City, New York: United Nations.
42. O'Connor, H. (2019, June 22). "Technologies of Control: We Have To Defend our Right of Refusal." *LSE Business Review*. https://blogs.lse.ac.uk/businessreview/2019/06/22/technologies-of-control-we-have-to-defend-our-right-of-refusal/.
43. This metaphor of "waves" is contested by those who see it as centering the contributions of white women and ignoring sustained organizing by women of color.
44. Bravo, L., C. Rufs, & D. Moyano. (2022). "Data Visualization for Non-oppression and Liberation: A Feminist Approach." *Diseña*, 21(21). https://doi.org/10.7764/disena.21.Article.2.
45. D'Ignazio, C. & L. F. Klein. (2020). *Data Feminism*. Cambridge, Massachusetts: The MIT Press.
46. Lee, V. R., D. R. Pimentel, R. Bhargava, & C. D'Ignazio. (2022). "Taking Data Feminism to School: A Synthesis and Review of Pre-Collegiate Data Science Education Projects." *British Journal of Educational Technology*, 53(5), 1096–1113,https://doi.org/10.1111/bjet.13251.
47. This work culminated in, and connected to, his Handmade Visualization Toolkit—a packaged supply of craft materials that could be used to make three dimensional versions of charts and graphs in public space. That work connects strongly with, and informed, my own work on data sculptures that was beginning at the same time.
48. Bird, J. & Y. Rogers. (2010). The pulse of tidy street: Measuring and publicly displaying domestic electricity consumption. In *workshop on energy awareness and conservation through pervasive applications (Pervasive 2010)* (Vol. 195). Helsinki, Finland.
49. While describing the data murals work I use collective pronouns to capture the collaborative nature of the work created by my wife Emily and myself. The ideation, process design, facilitation, design, and production of all the data murals are the results of our collaboration.
50. Head Start is a federal program in the US that offers a variety of services to low-income families, including health, community building, and childcare.
51. All these hands-on activities are documented for educators and facilitators as part of the Data Culture Project, available at the databasic.io website.
52. Bhargava, R., R. Kadouaki, E. Bhargava, G. Castro, & C. D'Ignazio. (2016). "Data Murals: Using the Arts to Build Data Literacy." *The Journal of Community Informatics*, 12(3). 197–216.
53. Ruvaga, L. (2014, August 6). "Kenyan Graffiti Artists Spray for Political, Social Change." *Voice of America*. https://www.voanews.com/a/kenyan-graffiti-artists-spray-for-political-social-change/1972955.html.
54. swift graffiti [@swiftgraffiti]. (2022, July 21). *Mural for @MarthaKarua The next Deputy president! Https://t.co/w4soE2GKvi* [Tweet]. Twitter. https://twitter.com/swiftgraffiti/status/1550048885795704833.
55. Offenhuber, D. (2019). "Data by Proxy—Material Traces as Autographic Visualizations." *arXiv:1907.05454 [Cs, Math]*. http://arxiv.org/abs/1907.05454.
56. Taylor, A. S., S. Lindley, T. Regan, D. Sweeney, V. Vlachokyriakos, L. Grainger, et al. (2015). *Data-in-Place: Thinking through the Relations Between Data and Community*, 2863–2872. https://doi.org/10.1145/2702123.2702558.
57. *Life Under Curfew—Social Justice Centres Working Group*. (n.d.). Retrieved July 19, 2023, from https://www.sjc.community/life-under-curfew.
58. Social Justice Centre Working Group. (2022). " Life under Curfew." Information Is Beautiful Awards, https://www.informationisbeautifulawards.com/showcase/5559-life-under-curfew.

REFERENCES 225

59. Data Murals Tackle Drug Abuse and Reproductive Health Challenges in Dar Es Salaam | Data Collaboratives for Local Impact. (n.d.). *Data Collaboratives for Local Impact.* Retrieved July 19, 2023, from https://dcli.co/impact-story/using-a-data-mural-to-fight-drug-abuse/.
60. Heath, R. (2015, September 21). "What Redlining Looks Like: City Life/Vida Urbana Takes to the Streets to Outline Housing Discrimination." *Jamaica Plain News*, https://www.jamaicaplainnews.com/2015/09/21/what-redlining-looks-like-city-lifevida-urban-takes-to-the-streets-to-outline-housing-discrimination/13749.
61. Houseman, H. (2015, September 22). "Artists and Activists Trace Boston's Historic Red Line on the Streets." *Hyperallergic*, https://hyperallergic.com/238667/artists-and-activists-trace-bostons-historic-red-line-on-the-streets/.

Chapter 3

1. Clayton, A. (2020, October 27). "How Eugenics Shaped Statistics." *Frontiers—Nautilus*, 92, https://nautil.us/how-eugenics-shaped-statistics-238014/.
2. Smith, A. (Director). (2019, October 10). *Eugenics: Science's Greatest Scandal.* BBC Four, https://www.bbc.co.uk/programmes/m0008zc7.
3. Villarosa, L. (2022, June 8). "The Long Shadow of Eugenics in America." *The New York Times*, https://www.nytimes.com/2022/06/08/magazine/eugenics-movement-america.html.
4. Harwell, D. & E.Dou. (2020, December 8). "Huawei Tested AI Software that could Recognize Uighur Minorities and Alert Police, Report Says." *The Washington Post*, https://www.washingtonpost.com/technology/2020/12/08/huawei-tested-ai-software-that-could-recognize-uighur-minorities-alert-police-report-says/.
5. Ziogas, A., A. Mokros, W. Kawohl, M. de Bardeci, I. Olbrich, B. Habermeyer, et al. (2023). "Deep Learning in the Identification of Electroencephalogram Sources Associated with Sexual Orientation." *Neuropsychobiology*, 82(4), 234–245, https://doi.org/10.1159/000530931.
6. Radke, H. (2022). *Butts: A Backstory.* Avid Reader Press/Simon & Schuster. New York, New York.
7. Felsenthal, J. (2012, January 25). "A Size 2 Is a Size 2 Is a Size 8." *Slate*, https://slate.com/culture/2012/01/clothing-sizes-getting-bigger-why-our-sizing-system-makes-no-sense.html.
8. Amandeep Gill [@gioasempre]. (2018, October 29). https://t.co/c1WSCCApqL [Tweet]. Twitter, https://twitter.com/gioasempre/status/1056959155498557441.
9. Davenport, T. H. & D. J. Patil. (2012, October 1). "Data Scientist: The Sexiest Job of the 21st Century." *Harvard Business Review*, https://hbr.org/2012/10/data-scientist-the-sexiest-job-of-the-21st-century.
10. Mayer-Schnberger, Viktor. (2013). *Big Data: A Revolution That Will Transform How We Live, Work and Think. Viktor Mayer-Schnberger and Kenneth Cukier.* London, GBR: John Murray Publishers.
11. Benjamin, Solomon, R., Bhuvaneswari, P. Rajan, & Manjunatha. (2007). "Bhoomi: 'E-Governance', or, an Anti-Politics Machine Necessary to Globalize Bangalore?' Collaborative for the Advancement of Studies in Urbanism through Mixed Media." *CASUM–m Working Paper*, https://casumm.files.wordpress.com/2008/09/bhoomi-e-governance.pdf.
12. Gutiérrez, J. D., S. Muñoz-Cadena, & M.Castellanos-Sánchez. (2023). *Sistemas de decisión automatizada en el sector público colombiano* [dataset]. Universidad del Rosario, https://doi.org/10.34848/YN1CRT.
13. Valkanova, N., S. Jorda, M. Tomitsch, & A. Vande Moere. (2013). "Reveal-it!: The Impact of a Social Visualization Projection on Public Awareness and Discourse." Proceedings of the SIGCHI Conference on Human Factors in Computing Systems, 3461–3470, http://dl.acm.org/citation.cfm?id=2466476.
14. Valkanova, N., R. Walter, A. Vande Moere, & J.Müller. (2014). "My Position: Sparking Civic Discourse by a Public Interactive Poll Visualization." Proceedings of the 17th ACM Conference on Computer Supported Cooperative Work & Social Computing, 1323–1332, https://doi.org/10.1145/2531602.2531639.
15. Chang, C. (2013). *Before I Die* (first edn). St. Martin's Griffin.
16. 117th Congress (2023). Data Science and Literacy Act of 2023.
17. Schauffler, M. (2021, July 26). "Building Blocks for Data Literacy." *Partners in Data Literacy*, https://partnersindataliteracy.com/2021/07/26/building-blocks/.
18. Levi-Strauss, C. & P.Wilcken. (2012). *Tristes Tropiques*, J. Weightman & D. Weightman, trans., revised ed. edn. Penguin Classics.
19. It is worth noting here that Levi-Strauss does not acknowledge the very similar type of subjugation in his own work on introducing his structure into the Indigenous populations he worked with.

226 REFERENCES

20. Hindman Sandra L.'s (1984) article "With Ink and Mortar: Christine de Pizan's 'Cite des Dames.'" in *Feminist Studies*, 10(3) (Autumn), 457 indicates that the source was from Phillip de Navarre, *Les Quatre ages de l'homme* and the anonymous *Libro di buoni costume*, respectively.
21. Ealey, S. (2022, January 12). "Literacy By Any Means Necessary: The History of Anti-Literacy Laws in the U.S. Oakland Literacy Coalition." https://oaklandliteracycoalition.org/literacy-by-any-means-necessary-the-history-of-anti-literacy-laws-in-the-u-s/.
22. Narrative of Proctor, Jenny. (2007). "Enslaved in Alabama, 1850–1865." In *The Making of African American Identity*, vol. I, 1500–1865. National Humanities Center.
23. His numerous books offer deeper thinking than the brief summaries I offer here, the 1968 *Pedagogy of the Oppressed* offering the best summary.
24. Walker, R. (2023). *Liberatory Computing Framework: Empowering High School Students to Mitigate Systemic Oppression through Data Activism*. Massachusetts Institute of Technology, https://hdl.handle.net/1721.1/151995.
25. Ouellette, K. (2022, September 21). "Empowering Cambridge Youth through Data Activism." *MIT News*, https://news.mit.edu/2022/empowering-cambridge-youth-through-data-activism-0921.
26. Downey pointed out to me more recently that he thinks he is paraphrasing someone, perhaps statistician John Tukey.
27. Harouni, H. (2015). "Toward a Political Economy of Mathematics Education." *Harvard Educational Review*, 85, 50–74, https://doi.org/10.17763/haer.85.1.2q580625188983p6.
28. Leonard, Alison E., Nicole A. Bannister, and Nikeetha Farfan D'Souza. (2020). "'(Non)Dance and (Non)Math People': Challenging Binary Disciplinary Identities in Education." *Research in Dance Education*, 0(0), 1–19, https://doi.org/10.1080/14647893.2020.1853692.
29. UNICEF Rwanda & National Institute of Statistics Rwanda. (2016). "Teachers' Discussion Guide for Reading Data with Children." UNICEF, https://www.unicef.org/rwanda/reports/teachers-discussion-guide-reading-data-children.
30. Bhaskaran, H., G. Kashyap, & H. Mishra. (2022). "Teaching Data Journalism: A Systematic Review." *Journalism Practice*, 0(0), 1–22, https://doi.org/10.1080/17512786.2022.2044888.
31. Oliver, J. C. &, T.McNeil. (2021). "Undergraduate Data Science Degrees Emphasize Computer Science and Statistics but Fall Short in Ethics Training and Domain-Specific Context." *Peer Journal of Computer Science*, 7, e441, https://doi.org/10.7717/peerj-cs.441.
32. National Academy of Sciences, National Academy of Engineering, and Institute of Medicine. (2007). *Rising Above the Gathering Storm: Energizing and Employing America for a Brighter Economic Future*. Washington, DC: The National Academies Press, https://doi.org/10.17226/11463.
33. Bhargava, Rahul. (2019). "Data Literacy." In *International Encyclopedia of Media Literacy Education*, R. Hobbs and P. Mihailidis (eds.). Hoboken, NJ: Wiley-Blackwell.
34. Gould, R. (2017). "Data Literacy Is Statistical Literacy." *Statistics Education Research Journal*, 16(1), 22–25.
35. Disclosure: I was briefly one of the members of the advisory board for Qlik's Data Literacy Project, from which this report emerged (after my time there); Accenture. (2020). *The Human Impact of Data Literacy* (Data Literacy Project), Qlik, https://www.accenture.com/_acnmedia/PDF-115/Accenture-Human-Impact-Data-Literacy-Latest.pdf.
36. Cairo, A. (2019, July 23). "Democratizing Visualization By Lowering the Barrier of Entry." *Nightingale*, https://medium.com/nightingale/democratizing-visualization-by-lowering-the-barrier-of-entry-c46d30a2ab41.
37. Learn more at https://elab.emerson.edu/research/projects/civic-data-ambassadors/.
38. Ramirez, R. (2020, December 3). "Obama Said 'Defund the Police' Is a Bad Slogan. This shouldn't Come as a Surprise." *Vox*, https://www.vox.com/2020/12/3/22150452/obama-defund-the-police-snappy-slogan.
39. Rashawn, R. (2020, June 19). *What Does "Defund the Police" Mean and Does it Have Merit?* Brookings Institute, https://www.brookings.edu/articles/what-does-defund-the-police-mean-and-does-it-have-merit/.
40. Barker, K., M. Baker, & A.Watkins. (2021, March 20). "In City after City, Police Mishandled Black Lives Matter Protests." *The New York Times*, https://www.nytimes.com/2021/03/20/us/protests-policing-george-floyd.html.
41. Ransby-Sprona, A. (2020, 4 July). "These Boxes Visualize Chicago's City Budget. The Tall Stack Is the $1.8 Billion Police Budget. Housing & Public Health each only Have One Box." #Defund-Police #DefundCPD https://t.co/7IEcYvjtd7 [Tweet]. Twitter, https://twitter.com/ashapoesis/status/1279489934576046085.

REFERENCES 227

42. In addition, I find the word "physicalization" impossible to pronounce in public talks and teaching.
43. Hutmacher, F. & C.Kuhbandner. (2018). "Long-Term Memory for Haptically Explored Objects: Fidelity, Durability, Incidental Encoding, and Cross-Modal Transfer." *Psychological Science*, 29(12), 2031–2038, https://doi.org/10.1177/0956797618803644.
44. Hall, P. A. & P.Dávila. (2022). *Critical Visualization: Rethinking the Representation of Data.* Bloomsbury, https://www.bloomsbury.com/us/critical-visualization-9781350077256/, 52–54.
45. Medrano, M. & G.Urton. (2018). "Toward the Decipherment of a Set of Mid-Colonial Khipus from the Santa Valley, Coastal Peru." *Ethnohistory*, 65(1), 1–23, https://doi.org/10.1215/00141801-4260638.
46. Ascher, M. (1995). "Models and Maps from the Marshall Islands: A Case in Ethnomathematics." *Historia Mathematica*, 22(4), 347–370, https://doi.org/10.1006/hmat.1995.1030.
47. Rojas, J. & J.Kamp. (2022). *Dream Play Build*. Island Press, https://islandpress.org/books/dream-play-build.
48. Bhargava, R. &, C.D'Ignazio. (2017, 10 June). *Data Sculptures as a Playful and Low-Tech Introduction to Working with Data*. Edinburgh, Scotland: Designing Interactive Systems, http://dataphys.org/workshops/dis17/wp-content/uploads/sites/5/2017/06/Data_Phys_2017_Workshop-rev2.pdf.
49. Youden, W. J. (1962). *Experimentation and Measurement*. Scholastic Book Services.
50. Huron, S., T. Nagel, L. Oehlberg, & W. Willett. (eds.) (2022). *Making with Data: Physical Design and Craft in a Data-Driven World* (first edn). A K Peters/CRC Press.
51. Bae, S. S., C. Zheng, M. E. West, E. Y.-L. Do, S. Huron, & D. A.Szafir. (2022). *Making Data Tangible: A Cross-disciplinary Design Space for Data Physicalization* (arXiv:2202.10520). arXiv, https://doi.org/10.48550/arXiv.2202.10520.
52. Zhao, J. & A. V.Moere. (2008). "Embodiment in Data Sculpture: A Model of the Physical Visualization of Information." Proceedings of the 3rd International Conference on Digital Interactive Media in Entertainment and Arts, 343–350, https://doi.org/10.1145/1413634.1413696.
53. Vande Moere, A. & S.Patel. (2010). "The Physical Visualization of Information: Designing Data Sculptures in an Educational Context." In *Visual Information Communication*, M. L. Huang, Q. V. Nguyen, & K. Zhang (eds.), 1–23. US: Springer, https://doi.org/10.1007/978-1-4419-0312-9_1.
54. Chandler, D. (2007). *Semiotics: The Basics* (second edn). Routledge.
55. Dumičić, Ž., K. Thoring, H. Klöckner, & G. Joost. (2022). "Design Elements in Data Physicalization: A Systematic Literature Review." DRS Biennial Conference Series, https://dl.designresearchsociety.org/drs-conference-papers/drs2022/researchpapers/234.
56. Available at, http://www.aviz.fr/phys.
57. Jansen, Y. & K. Hornbaek. (2016). "A Psychophysical Investigation of Size as a Physical Variable." *Visualization and Computer Graphics, IEEE Transactions On*, 22(1), 479–488.
58. Sauvé, K., S. Bakker, & S.Houben. (2020). "Econundrum: Visualizing the Climate Impact of Dietary Choice through a Shared Data Sculpture." Proceedings of the 2020 ACM Designing Interactive Systems Conference, 1287–1300, https://doi.org/10.1145/3357236.3395509.
59. Kahn, P., H. Dubberly, & D.Rodighiero. (2022). "COVIC: Collecting Visualizations of COVID-19 to Outline a Space of Possibilities." *Design Issues*, 38(4), 44–62, https://doi.org/10.1162/desi_a_00697.
60. Perovich, L. J., M. Alper, & C.Cleveland. (2022). "'Self-Quaranteens' Process COVID-19: Understanding Information Visualization Language in Memes." Proceedings of the ACM on Human-Computer Interaction, 6 (47), 1–20, https://doi.org/10.1145/3512894.
61. This was in stark contrast to President Trump's rambling, uninformative, intentionally misleading daily White House updates. Cuomo felt to many like a voice of reason and empathy within the information storm. It is important to note that Cuomo's status as a well-reasoned and respected nationwide leader was short lived, eradicated by revelations of multiple accusations of sexual misconduct and harassment.
62. Ellison, S. & Terris, B. (2020, March 30). "Andrew Cuomo during the Covid-19 Crisis Is the same as Ever, with One Big Difference: People Like Him." *Washington Post*, https://www.washingtonpost.com/lifestyle/style/andrew-cuomo-during-the-covid-19-crisis-is-the-same-as-ever-with-one-big-difference-people-like-him/2020/03/28/11a89a0a-6fd7-11ea-b148-e4ce3fbd85b5_story.html.
63. WNYT NewsChannel 13 (Director). (2020, June 29). *Cuomo Unveils Coronavirus Mountain*, https://www.youtube.com/watch?v=hiVBlaUEKFE.

228 REFERENCES

64. Campbell, J. & Spector, J. (2020, June 29). "Cuomo Unveils a Giant Foam Mountain Depicting COVID-19 Climb in NY. Social Media Took over." *Democrat and Chronicle*, https://www.democratandchronicle.com/story/news/politics/albany/2020/06/29/andrew-cuomo-unveils-foam-mountain-visualize-nys-coivd-19-fight/3278852001/.

65. López García, I. & Hornecker, E. (2021). "Scaling Data Physicalization—How Does Size Influence Experience?" Proceedings of the Fifteenth International Conference on Tangible, Embedded, and Embodied Interaction, 1–14, https://doi.org/10.1145/3430524.3440627.

66. Project Bread. (2022). 2022 Annual Report, https://projectbread.org/uploads/attachments/clfveuez403dr1r9hgu5x00zx-ar-2022-digital-compressed.pdf.

67. Furneaux, R. (1974). *William Wilberforce* (first edn). Hamish Hamilton. https://www.abebooks.com/first-edition/William-Wilberforce-Robin-Furneaux-Hamish-Hamilton/11440304269/bd.

68. History of the Rise, Progress, and Accomplishment of the Abolition of the African Slave Trade (1808)

69. The British Museum & BBC. (n.d.). "The 'Brookes' Slave Ship Model." *BBC*. Retrieved October 30, 2023, from https://www.bbc.co.uk/ahistoryoftheworld/objects/TrVw1QSnSKyRI59LC2csxg.

70. Papert, S. (1980). *Mindstorms: Children, Computers, and Powerful Ideas*. Basic Books.

71. Dangol, A. & S.Dasgupta. (2023). "Constructionist Approaches to Critical Data Literacy: A Review." Proceedings of the 22nd Annual ACM Interaction Design and Children Conference, 112–123, https://doi.org/10.1145/3585088.3589367.

72. Huron, S., S. Carpendale, J. Boy, & J.-D. Fekete. (2016). "Using VisKit: A Manual for Running a Constructive Visualization Workshop." Pedagogy of Data Visualization Workshop at IEEE VIS 2016, https://hal.inria.fr/hal-01384388/; Huron, S. & W.Willet. (2021). "Visualizations as Data Input?" *IEEE altVIS*, https://altvis.github.io/papers/willett.pdf.

73. Huron, S., Y. Jansen, & S. Carpendale. (2014). "Constructing Visual Representations: Investigating the Use of Tangible Tokens," EEE Transactions on Visualization and Computer Graphics, Institute of Electrical and Electronics Engineers (IEEE), Transactions on Visualization and Computer Graphics, 20(12), 1, https://doi.org/10.1109/TVCG.2014.2346292.

74. Lupton, D. (2017). "Feeling your Data: Touch and Making Sense of Personal Digital Data." *New Media & Society*, 19(10), 1599–1614, https://doi.org/10.1177/1461444817717515; Lallemand, C. & M. Oomen. (2022). "The Candy Workshop: Supporting Rich Sensory Modalities in Constructive Data Physicalization." Extended Abstracts of the 2022 CHI Conference on Human Factors in Computing Systems, 1–7, https://doi.org/10.1145/3491101.3519648.

75. *Data as Culture—The ODI's Art Programme*. (n.d.). ODI Open Data Institute. Retrieved May 18, 2023, from https://culture.theodi.org/.

76. Dragicevic, P. & Y.Jansen. (n.d.). "List of Physical Visualizations and Related Artifacts." *Data Physicalization Wiki*. Retrieved August 11, 2023, from http://dataphys.org/list/.

77. Confessore, N., S. Cohen, & K. Yourish. (2015, October 10). "The Families Funding the 2016 Presidential Election." *The New York Times*, https://www.nytimes.com/interactive/2015/10/11/us/politics/2016-presidential-election-super-pac-donors.html.

78. "The most Popular Board Games of all Time." (2023, March 20). *The Baltimore Sun*, https://www.baltimoresun.com/entertainment/comics/games/sns-the-most-popular-board-games-of-all-time-20230320-52cwjqgdhvcljdjqzegonn5hv4-photogallery.html.

Chapter 4

1. Bourdieu, P. (1979). "Symbolic Power." *Critique of Anthropology*, 4(13–14), 77–85, https://doi.org/10.1177/0308275X7900401307.

2. Schmidt, S. (2022, July 3). "How Green became the Color of Abortion Rights." *Washington Post*. https://www.washingtonpost.com/world/interactive/2022/abortion-green-roe-wade-argentina/.

3. Dedman, R. (2023). "The Politicisation of Palestinian Embroidery Since 1948." In *Dangerous Bodies: New Global Perspectives on Fashion and Transgression*,R. Mahawatte & J. Willson (eds.), 97–116. Springer International Publishing. https://doi.org/10.1007/978-3-031-06208-7_6.

4. The spelling of khanga is a transliteration, and could also be spelled *kanga*. I have chosen this spelling because it is the one used by the groups who ran the project in their English language writing.

5. Andersen, M. (2018, February 22). "The Secret, Subversive + Revolutionary Language of the East African Kanga." *Eye on Design*, https://eyeondesign.aiga.org/the-secret-subversive-revolutionary-language-of-the-east-african-kanga/.

6. Beck, R. M. (2001). "Ambiguous Signs: The Role of the Kanga as a Medium of Communication." *Swahili Forum*, 68, 157–169.

REFERENCES   229

7. Data Zetu was a three-year development program implemented by the International Research & Exchanges Board (IREX) in partnership with several Tanzanian CSOs, including the Tanzania Bora Initiative (TBI). The project was funded by PEPFAR in partnership with the Millennium Challenge Corporation.

8. Katuli, M. (2019, January 2). "Young Artists Use Fashion and Data To Promote Dialog on Sexual Health." *Data Zetu*, https://medium.com/data-zetu/young-artists-use-fashion-and-data-to-promote-dialog-on-sexual-health-517429662ec2.

9. Katuli, M. (2019, January 2). "Young Artists Use Fashion and Data To Promote Dialog on Sexual Health." *Data Zetu*, https://medium.com/data-zetu/young-artists-use-fashion-and-data-to-promote-dialog-on-sexual-health-517429662ec2.

10. Gebre, E. H. (2018). "Young Adults' Understanding and Use of Data: Insights for Fostering Secondary School Students' Data Literacy." *Canadian Journal of Science, Mathematics and Technology Education*, 18(4), 330–341. https://doi.org/10.1007/s42330-018-0034-z.

11. Schoffelen, J., S. Claes, L. Huybrechts, S. Martens, A. Chua, & A. V.Moere. (2015). "Visualising Things. Perspectives on How To Make Things Public through Visualisation." *CoDesign*, 11(3–4), 179–192, https://doi.org/10.1080/15710882.2015.1081240.

12. Venturini, T. (2012). "Building on Faults: How To Represent Controversies with Digital Methods." *Public Understanding of Science*, 21(7), 796–812. https://doi.org/10.1177/0963662510387558.

13. Sprague, David, & Melanie Tory. (2012). "Exploring how and why people use visualizations in casual contexts: Modeling user goals and regulated motivations." *Information Visualization*, 11(2), 106–123.

14. Hullman, J., E. Adar, & P.Shah. (2011). "Benefitting InfoVis with Visual Difficulties." *IEEE Transactions on Visualization and Computer Graphics*, 17(12), 2213–2222. https://doi.org/10.1109/TVCG.2011.175.

15. Tufte, Edward R. (2001). "Graphical Excellence." In *The Visual Display of Quantitative Information*, second edn. Cheshire, Conn: Graphics Press.

16. Cairo, A. (2023). *The Art of Insight: How Great Visualization Designers Think* (first edn). Wiley, Hoboken, New Jersey.

17. Harding, S. (2007). "Feminist Standpoints." In Sharlene Nagy Hesse-Biber, *Handbook of Feminist Research: Theory and Praxis*, 45–69. SAGE Publications.

18. Hall, P. A. & Dávila, P. (2022). *Critical Visualization: Rethinking the Representation of Data.* Bloomsbury.

19. Tufte, E. R. (2001). *The Visual Display of Quantitative Information* (second edn). Graphics Press.

20. Dick, M. (2020). *The Infographic*, 19. MIT Press.

21. Borneman, E. (2020). *Data Visualizations for Perspective Shifts and Communal Cohesion.* MIT.

22. Within this book I will be using the terms "chart" and "graph" interchangeably. While I believe a technical distinction can be made, in my experience with popular vernacular use they are equivalent.

23. Bertin, Jacques. (1967). *Semiology of Graphics: Diagrams, Networks, Maps.* Esri Press, Redlands, California.

24. Almost forty years later Mackinlay's goal, to automate the creation of "effective" charts based on the data with a software analysis tool, is still being pursued today. Tools like Table, Excel, and Google Sheets have in the last few years added automatic chart creation based on tabular data, exactly as Mackinlay hoped.

25. Amara, J., P. Kaur, M. Owonibi, & B.Bouaziz. (n.d.). Convolutional Neural Network Based Chart Image Classification.https://doi.org/10.1109/TVCG.2012.262.

26. Vougiouklis, P., L. Carr, & E.Simperl. (2020). "Pie Chart or Pizza: Identifying Chart Types and Their Virality on Twitter." Proceedings of the International AAAI Conference on Web and Social Media, 14, 694–704, https://doi.org/10.1609/icwsm.v14i1.7335.

27. Ajani, K., E. Lee, C. Xiong, C. N. Knaflic, W. Kemper, & S.Franconeri. (2022). "Declutter and Focus: Empirically Evaluating Design Guidelines for Effective Data Communication." *IEEE Transactions on Visualization and Computer Graphics*, 28(10), 3351–3364, https://doi.org/10.1109/TVCG.2021.3068337.

28. Wood, J., P. Isenberg, T. Isenberg, J. Dykes, N. Boukhelifa, & A. Slingsby. (2012). "Sketchy Rendering for Information Visualization." *IEEE Transactions on Visualization and Computer Graphics*, 18(12), 2749-2758. https://doi.org/10.1109/TVCG.2012.262.

29. Bateman, S., R. L. Mandryk, C. Gutwin, A. Genest, D. McDine, & C.Brooks. (2010). "Useful Junk? The Effects of Visual Embellishment on Comprehension and Memorability of Charts."

230 REFERENCES

Proceedings of the SIGCHI Conference on Human Factors in Computing Systems, 2573–2582, https://doi.org/10.1145/1753326.1753716.

30. Borkin, M. A., Z. Bylinskii, N. W. Kim, C. M. Bainbridge, C. S. Yeh, D. Borkin, et al. (2016). "Beyond Memorability: Visualization Recognition and Recall." *IEEE Transactions on Visualization and Computer Graphics*, 22(1), 10.

31. Pandey, A. V., K. Rall, M. L. Satterthwaite, O. Nov, & E.Bertini. (2015). "How Deceptive are Deceptive Visualizations? An Empirical Analysis of Common Distortion Techniques." Proceedings of the 33rd Annual ACM Conference on Human Factors in Computing Systems, 1469–1478, https://doi.org/10.1145/2702123.2702608.

32. Zhang, S., P. R. Heck, M. N. Meyer, C. F. Chabris, D. G. Goldstein, & J. M. Hofman. (2022). *An Illusion of Predictability in Scientific Results* [Preprint]. SocArXiv, https://doi.org/10.31235/osf.io/5tcgs.

33. Weissgerber, T. L., N. M. Milic, S. J. Winham, & V. D. Garovic. (2015). "Beyond Bar and Line Graphs: Time for a New Data Presentation Paradigm." *PLOS Biology*, 13(4), e1002128, https://doi.org/10.1371/journal.pbio.1002128.

34. Driessen, J. E. P., D. A. C. Vos, I. Smeets, & C. J.Albers. (2022). "Misleading Graphs in Context: Less Misleading than Expected." *PLOS ONE*, 17(6), e0265823, https://doi.org/10.1371/journal.pone.0265823.

35. Cleveland, W. and R. McGil. (1984). "Graphical Perception: Theory, Experimentation, and Application to the Development of Graphical Methods." *Journal of the American Statistical Association*, 79(387), 531–554.

36. Peck, Evan M., Sofia E. Ayuso, and Omar El-Etr. (2019). "Data Is Personal: Attitudes and Perceptions of Data Visualization in Rural Pennsylvania." In Proceedings of the 2019 CHI Conference on Human Factors in Computing Systems—CHI '19, 1–12. Glasgow, Scotland UK: ACM Press, https://doi.org/10.1145/3290605.3300474.

37. He, H. A., J. Walny, S. Thoma, S. Carpendale, & W.Willett. (2023). *Enthusiastic and Grounded, Avoidant and Cautious: Understanding Public Receptivity to Data and Visualizations* [Preprint]. Open Science Framework, https://doi.org/10.31219/osf.io/mpq32.

38. Watson, A. (2022). "Youth, Zines, and Music Scenes." In *The Bloomsbury Handbook of Popular Music and Youth Culture*, A. Bennett (ed.), 411–426. USA: Bloomsbury Publishing.

39. Fernando, C. (2021, July 31). "'Zines' Have Deep Roots in Marginalized Communities. Zine-Makers Worry those Origins Are Being Forgotten." *USA TODAY*, https://www.usatoday.com/story/news/nation/2021/07/31/zines-rooted-protest-culture-marginalized-communities/8031819002/.

40. Ketchum, A. & N.Morena. (2022). "AI, Big Data, and Surveillance Zines as Forms of Community Healthcare." *Feminist Media Studies*, 0(0), 1–17, https://doi.org/10.1080/14680777.2022.2149598.

41. Onuoha, M. & D.Nucera. (2018). *A People's Guide to AI*. Allied Media Projects, Detroit, Michigan.

42. Takahashi, A., S. Wang, & C.Li. (2019). *Oh, the Places Your Data Will Go!* [Zine]. Free Radicals, https://freerads.org/2020/09/18/oh-the-places-your-data-will-go.

43. Wilson, Tess. (2017, September 28). "Civic Data Zine Camp." Young Adult Library Services Association, https://yalsa.ala.org/blog/2017/09/28/civic-data-zine-camp/.

44. Gradeck, B. & E.Tutt. (2017, October 19). "Pittsburgh's Data Day: Using Civic Data to Spark Hands-on Community Engagement." *Living Cities*, https://livingcities.org/blog/pittsburghs-data-day-using-civic-data-to-spark-hands-on-community-engagement/.

45. Gonick, L. & W.Smith. (1993). *The Cartoon Guide to Statistics* (first edn). HarperPerennial.

46. London, K., S. Farnham, & M.Lahav. (2014). "Data-Driven Cartoons: A Procedural Experiment in Provocation and Civic Participation." *Computation + Journalism Symposium in New York*.

47. Bhargava, R., A. Brea, L. Perovich, & J. Jinson. (2022). "Data Theatre as an Entry Point to Data Literacy." *Educational Technology & Society*, 25(4), https://researchportal.helsinki.fi/files/258336131/ETS_25_4_08.pdf.

48. Lea, G. W. & G.Belliveau. (eds.). (n.d.). *Research-Based Theatre*. Intellect Books. Retrieved July 20, 2023, from https://www.intellectbooks.com/research-based-theatre; Shigematsu, T., G. Lea, C. Cook, & G.Belliveau. (2022). "A Spotlight on Research-Based Theatre." *Learning Landscapes*, 15, 349–365. https://doi.org/10.1656112522.

49. Gray, R. E. & C.Sinding. (2002). *Standing Ovation: Performing Social Science Research about Cancer*. AltaMira Press.

50. Bogart, A. & T.Landau. (2004). *The Viewpoints Book: A Practical Guide to Viewpoints and Composition* (first edn). Theatre Communications Group.

REFERENCES 231

51. Snyder-Young, Dani, Michael Arnold Mages, Rahul Bhargava, Jonathan Carr, Laura Perovich, Victor Talmadge, Oliver Wason, Moira Zellner, Angelique C-Dina, Ren Birnholz, and et al. (2024). *"Viewpoints/Points of View: Building a Transdisciplinary Data Theatre Collaboration in Six Scenes"*. *Arts*, 13(1), 37. https://doi.org/10.3390/arts13010037

52. Damásio, António (1994). *Descartes' Error: Emotion, Reason, and the Human Brain*. Putnam, ISBN 0-399-13,894-3.

53. Lowenfeld, V. (1957). *Creative and Mental Growth* (third edn). Macmillan.

54. Posner, M. I., M. J. Nissen, & R. M.Klein. (1976). "Visual Dominance: An Information-Processing Account of its Origins and Significance." *Psychological Review*, 83(2), 157–171, https://doi.org/10.1037/0033-295X.83.2.157.

55. Posner, M. I. (1967). "Characteristics of Visual and Kinesthetic Memory Codes." *Journal of Experimental Psychology*, 75(1), 103–107, https://doi.org/10.1037/h0024911.

56. Höök, K. (2018). *Designing with the Body: Somaesthetic Interaction Design*. MIT Press.

57. Kelly, Spencer D., Dale J. Barr, R. Breckinridge Church, & Katheryn Lynch. (1999). "Offering a hand to pragmatic understanding: the role of speech and gesture in comprehension and memory." *Journal of Memory and Language*, 40, 577-592.

58. Yap, A. C. (2016). "How to Teach Kids Empathy Through Dance." *The Atlantic*, https://www.theatlantic.com/education/archive/2016/01/learning-empathy-through-dance/426498/.

59. Malinverni, L. & N. Pares. (2014). "Learning of Abstract Concepts through Full-Body Interaction: A Systematic Review." *Educational Technology & Society*, 17, 100–116.

60. Abrahamson, D. & R.Lindgren. (2014). "Embodiment and Embodied Design." In *The Cambridge Handbook of the Learning Sciences*, second edn, 358–376. Cambridge University Press, https://doi.org/10.1017/CBO9781139519526.022; Malinverni, L. & N. Pares. (2014). "Learning of Abstract Concepts through Full-Body Interaction: A Systematic Review." *Journal of Educational Technology & Society*, 17(4), 100–116.

61. Papert was likely building on Freud's concept of "egosyntonicity," used to describe values and feelings that are in harmony with one's ego or self-image.

62. Papert, S. (1980). *Mindstorms: Children, Computers, and Powerful Ideas*, 63. Basic Books.

63. Maxwell, L. E. & G. W.Evans. (2002). "Museums as Learning Settings: The Importance of the Physical Environment." *Journal of Museum Education*, 27(1), 3–7, https://doi.org/10.1080/10598650.2002.11510454.

64. Matuk, C., R. Vacca, A. Amato, M. Silander, K. DesPortes, P. J. Woods, et al. (2023). "Promoting Students' Informal Inferential Reasoning through Arts-Integrated Data Literacy Education." *Information and Learning Sciences*, 125(3-4), 163-189. https://doi.org/10.1108/ILS-07-2023-0088.

65. Shapiro, L. & S.Spaulding. (n.d.). "Embodied Cognition." In *The Stanford Encyclopedia of Philosophy*, E. N. Zalta (ed.), (Winter 2021 edn). Retrieved 20 July, 2023, from https://plato.stanford.edu/archives/win2021/entries/embodied-cognition/.

66. Lakoff, George, and Mark Johnson. (1980). *Metaphors We Live By*. Chicago: University of Chicago Press.

67. Goldin-Meadow, S. (2011). "Learning through Gesture." *WIREs Cognitive Science*, 2(6), 595–607, https://doi.org/10.1002/wcs.132.

68. McNeill, D. (2019). *Gesture and Thought*. University of Chicago Press, https://mcneilllab.uchicago.edu/pdfs/gesture.thought.fundamentals.pdf.

69. Lewis, G. M. (n.d.). "Maps, Mapmaking, and Map Use by Native North Americans." In *Cartography in the Traditional African, American, Arctic, Australian, and Pacific Societies*, vol. 2, D. Woodward & G. M. Lewis (eds.), University of Chicago Press.; Smethurst, G. (1774). *A Narrative of an Extraordinary Escape: Out of the Hands of the Indians, in the Gulph of St. Lawrence*. Kessinger Publishing, Whitefish, MT.

70. Sommer, S., & J. L. Polman. (2017). Embodied activities as entry points for science data literacy. In B. K. Smith, M. Borge, E. Mercier, & K. Y. Lim, *Making a Difference: Prioritizing Equity and Access in CSCL, 12th International Conference on Computer Supported Collaborative Learning, Volume 2* (pp. 849-850). Philadelphia: International Society of the Learning Sciences.

71. Barab, S. A., M. Cherkes-Julkowski, R. Swenson, S. Garrett, R. E. Shaw, & M.Young,. (1999). "Principles of Self-Organization: Learning as Participation in Autocatakinetic Systems." *Journal of the Learning Sciences*, 8(3–4), 349–390, https://doi.org/10.1080/10508406.1999.9672074.

72. Joiner, B. L. (1975). "Living Histograms." *International Statistical Review/Revue Internationale de Statistique*, 43(3), 339–340, https://doi.org/10.2307/1403117.

73. Joiner, B. L. (1975). "Living Histograms." *International Statistical Review/Revue Internationale de Statistique*, 43(3), 339–340, https://doi.org/10.2307/1403117.

232 REFERENCES

74. Solomon, F., L. Vogelstein, C. Brady, R. Steinberg, C. Thomas, D. Champion, et al. (2021). Embodying STEM: Learning at the intersection of Dance and STEM. In de Vries, E., Hod, Y., & Ahn, J. (Eds.), *Proceedings of the 15th International Conference of the Learning Sciences - ICLS 2021* (pp. 819-826). International Society of the Learning Sciences, Bochum, Germany.

75. Solomon, F., L. Vogelstein, C. Brady, R. Steinberg, C. Thomas, D. Champion, et al. (2021). Embodying STEM: Learning at the intersection of Dance and STEM. In de Vries, E., Hod, Y., & Ahn, J. (Eds.), *Proceedings of the 15th International Conference of the Learning Sciences - ICLS 2021* (pp. 819-826). International Society of the Learning Sciences, Bochum, Germany.

76. Sommer, S., & J. L. Polman. (2017). Embodied activities as entry points for science data literacy. In B. K. Smith, Borge M., Mercier E., & Lim K. Y., *Making a Difference: Prioritizing Equity and Access in CSCL, 12th International Conference on Computer Supported Collaborative Learning, Volume 2* (pp. 849-850). Philadelphia: International Society of the Learning Sciences.

77. Desportes, K., R. Vacca, M. Tes, P. Woods, C. Matuk, A. Amato, et al. (2022). "Dancing With data: Embodying the numerical and humanistic sides of data." *Proceedings of the 16th International Conference of the Learning Sciences-ICLS 2022* (pp. 305-312). International Society of the Learning Sciences.

78. Disclosure: I was an advisor on the National Science Foundation (NSF)-backed project run by the New York University (NYU) team in the public school system.

79. The US has a government run social security system, but has pushed the bulk of retirement planning to be privatized via employment-associated offerings. American workers are often provided retirement accounts by their employer, allowing them to invest in the stock market as the vehicle for retirement savings. The companies sometimes contribute limited matching funds in order to encourage this behavior. It is very much up to the individual young person to make the decision to start saving for retirement at a young age, which is not very high on most twenty-somethings to-do list for their newly earned money.

80. However, young readers should definitely look into retirement planning and insurance as soon as possible. I'm the son of a banker, and my grandfather-in-law worked for Prudential as an insurance salesman for decades, so perhaps I am a bit biased.

81. Lee, V. R., M. H. Wilkerson, & K. Lanouette. (2021). "A Call for a Humanistic Stance Toward K–12 Data Science Education." *Educational Researcher*, 50(9), 664–672, https://doi.org/10.3102/0013189X211048810.

82. Freire, P. (1968). *Pedagogy of the Oppressed*. Theatre Communications Group.

83. Boal, A. (1974). *Theatre of the Oppressed*, C. A. McBride (Trans.). Theatre Communications Group. Policy Press.

84. Jupp Kina, V. & K. C.Fernandes. (2017). "Augusto Boal's Theatre of the Oppressed: Democatising Art for Social Transformation." *Critical and Radical Social Work*, 5(2), 241–252, https://doi.org/10.1332/204986017X14951776937239.

85. Mumford, M. (2008). *Bertolt Brecht*. Routledge, https://doi.org/10.4324/9780203000991.

86. Styan, J. L. (1981). *Modern Drama in Theory and Practice*, vol. 3, *Expressionism and Epic Theatre*. Cambridge University Press.

87. Salas, R., K. Steele, A. Lin, C. Loe, L. Gauna, & P. Jafar-Nejad. (2013). "Playback Theatre as a Tool to Enhance Communication in Medical Education." *Medical Education Online*, 18, 22622, https://doi.org/10.3402/meo.v18i0.22622.

88. Dirnstorfer, A. & N. B. Saud. (2020). "A Stage for the Unknown? Reconciling Postwar Communities through Theatre-Facilitated Dialogue." *International Journal of Transitional Justice*, 14(1), 122–141, https://doi.org/10.1093/ijtj/ijz038.

89. Moran, G. S. & U.Alon. (2011). "Playback Theatre and Recovery in Mental Health: Preliminary Evidence." *The Arts in Psychotherapy*, 38(5), 318–324, https://doi.org/10.1016/j.aip.2011.09.002

90. Edelbi, K. (2020). *Using Playback Theater with Adolescents in Refugee Camps in Palestine to Tell Their Stories* [Doctor of Philosophy]. Lesley University.

91. Catchesides, K. (2020). "Verbatim Theatre." *Stockroom*, https://www.stockroom.co.uk/old-pages/verbatim-theatre/.

92. Farmer, D. (2014, December 22). "Image Theatre." *Drama Resource*, https://dramaresource.com/image-theatre/.

93. Sommer, D. (2014). *The Work of Art in the World: Civic Agency and Public Humanities*. Duke University Press.

94. Cohen-Cruz, J. & M.Schutzman. (eds.). (2006). *A Boal Companion: Dialogues on Theatre and Cultural Politics*. Routledge.

95. Chan, P. (2011). *Waiting for Godot in New Orleans: A Field Guide*. Badlands Unlimited.

REFERENCES   233

96. Hurtienne, J. (2018). "Possibilities of Human Data Embodiment: 100% City." Position Paper for the Workshop: Towards a Design Language for Data Physicalization at IEEE VIS, 6.
97. Fragapane, F. (2017, April 26). "Data Visualization and Theatre: A Story of Mutualism." *Medium*, https://medium.com/@frcfr/data-visualization-and-theatre-a-story-of-mutualism-5e199009cf4b.
98. Thorp, J. (2015, July 29). "A Sort of Joy." *Memo (Random)*, https://medium.com/memo-random/a-sort-of-joy-1d9d5ff02ac9.

Chapter 5

1. Science for All. (2019, November 20). Landmark Stories, https://www.youtube.com/watch?v=2e-j9nVhaW8.
2. Kaufmann, D., N. Hamidi, K. Palawat, & M. Ramirez-Andreotta. (2021). "Ripple Effect: Communicating Water Quality Data through Sonic Vibrations." Proceedings of the 13th Conference on Creativity and Cognition, 1–7, https://doi.org/10.1145/3450741.3464947.
3. Kaufmann, D. B., K. Palawat, S. Sandhaus, S. Buxner, E. McMahon, &, M. D. Ramírez-Andreotta. (2023). "Communicating Environmental Data through Art: The Role of Emotion and Memory in Evoking Environmental Action." *Humanities and Social Sciences Communications*, 10(1), article 1, https://doi.org/10.1057/s41599-023-02459-3.
4. Gardner, H. E. (2008). *Multiple Intelligences: New Horizons in Theory and Practice*. Basic Books. New York, New York.
5. Kolb, D. A. (1984). *Experiential Learning*. Prentice Hall. Hoboken, NJ.
6. Heron, J. & P. Reason. (2008). "Extending Epistemology within a Co-Operative Inquiry." In *The Sage Handbook of Action Research: Participative Inquiry and Practice*, 366–380. Sage. Thousand Oaks, CA.
7. Joiner, B. L. (1975). "Living Histograms." *International Statistical Review/Revue Internationale de Statistique*, 43(3), 339–340, https://doi.org/10.2307/1403117.
8. Turkle, S. & S. Papert. (1990). "Epistemological Pluralism: Styles and Voices within the Computer Culture." *Signs*, 16(1), 128–157.
9. Kramer, G., B. Walker, T. Bonebright, P. Cook, J. H. Flowers, N. Miner, et al. (2010.). "Sonification Report: Status of the Field and Research Agenda." Faculty Publications, Department of Psychology, 444.
10. Zanella, A., C. M. Harrison, S. Lenzi, J. Cooke, P. Damsma, & S. W. Fleming. (2022). "Sonification and Sound Design for Astronomy Research, Education and Public Engagement." *Nature Astronomy*, 6(11), article 11, https://doi.org/10.1038/s41550-022-01721-z.
11. Wanda Diaz (2016). *Merced: How a Blind Astronomer Found a Way to Hear the Stars*, https://www.ted.com/talks/wanda_diaz_merced_how_a_blind_astronomer_found_a_way_to_hear_the_stars.
12. Hermann, T., A. Hunt, & J. G. Neuhoff. (eds.). (2011). *The Sonification Handbook* (first edn). Logos Verlag Bersin. Berlin, Germany, https://sonification.de/handbook/; Kramer, G. (1994). *Auditory Display: Sonification, Audification, And Auditory Interfaces*. CRC Press. Boca Raton, FL.
13. Neuhoff, J. G. (2019, June). "Is Sonification Doomed To Fail?" International Conference on Auditory Display (ICAD), http://hdl.handle.net/1853/61531.
14. Sheikh, K. (2019, November 9). "This Is What Climate Change Sounds Like." *The New York Times*, https://www.nytimes.com/2019/11/09/science/climate-change-music-sound.html.
15. McGrory, H. (2019, March 5). "An Intro to TwoTone." *Datavized*, https://medium.com/@datavized/an-intro-to-twotone-7e10c7447a5d.
16. Lenzi, S., P. Ciuccarelli, H. Liu, & Y. Hua. (2021). "Data Sonification Archive," https://sonification.design/.
17. Lenzi, S. & P. Ciuccarelli. (2020). "Intentionality and Design in the Data Sonification of Social Issues." *Big Data & Society*, 7, 205395172094460, https://doi.org/10.1177/2053951720944603.
18. Cox, A. (2010, February 26). "Fractions of a Second: An Olympic Musical." *New York Times*, https://archive.nytimes.com/www.nytimes.com/interactive/2010/02/26/sports/olympics/20100226-olysymphony.html.
19. Smith, A. (2019, March 15). "Sonification: Turning the Yield Curve into Music." *Financial Times*, https://www.ft.com/content/80269930-40c3-11e9-b896-fe36ec32aece.
20. Quick, M. & D. Geere. (n.d.). "Loud Numbers." Retrieved December 24, 2023, from https://open.spotify.com/show/1tQOPau5LfZBhjnlFW4KeD.

234 REFERENCES

21. Backer, S. (2020, December 1). "Hear the Blind Spot: Visualizing Data for Those Who Can't See, *Nightingale*. https://medium.com/nightingale/hear-the-blind-spot-visualizing-data-for-those-who-cant-see-b370e6ec0b9e.

22. Guttman, S. E., L. A. Gilroy, & R. Blake. (2005). "Hearing what the Eyes See: Auditory Encoding of Visual Temporal Sequences." *Psychol. Sci.* 16, 228–235.

23. Adcock, B. (2022, January 22). "'I've Got this Little Extra Strength': The Rare, Intense World of a Super-Smeller." *The Guardian*, https://www.theguardian.com/society/2022/jan/23/ive-got-this-little-extra-strength-the-rare-intense-world-of-a-super-smeller.

24. Frumin, I., O. Perl, Y. Endevelt-Shapira, A. Eisen, N. Eshel, I. Heller, et al. (2015). "A Social Chemosignaling Function for Human Handshaking." *Elife*, 4, e05154, https://elifesciences.org/articles/05154.

25. Olsson, M. J., J. N. Lundström, B. A. Kimball, A. R. Gordon, B. Karshikoff, N. Hosseini, et al. (2014). "The Scent of Disease: Human Body Odor Contains an Early Chemosensory Cue of Sickness." *Psychological Science*, 25(3), 817–823, https://doi.org/10.1177/0956797613515681.

26. Ackerl, K., M. Atzmueller, & K. Grammer. (2002). "The Scent of Fear." *Neuro Endocrinology Letters*, 23(2), 79–84.

27. Cerda-Molina, A. L., L. Hernández-López, C. E. de la O, R. Chavira-Ramírez, & R. Mondragón-Ceballos. (2013). "Changes in Men's Salivary Testosterone and Cortisol Levels, and in Sexual Desire after Smelling Female Axillary and Vulvar Scents." *Frontiers in Endocrinology*, 4, 159, https://doi.org/10.3389/fendo.2013.00159.

28. Maner, J. K. & J. K. McNulty. (2013). "Attunement to the fertility status of same-sex rivals: Women's testosterone responses to olfactory ovulation cues." *Evolution and Human Behavior*, 34(6), 412–418, https://doi.org/10.1016/j.evolhumbehav.2013.07.005; Havlíček, J., R. Dvořáková, L. Bartoš, & J. Flegr. (2006). "Non-Advertized does not Mean Concealed: Body Odour Changes across the Human Menstrual Cycle." *Ethology*, 112(1), 81–90, https://doi.org/10.1111/j.1439-0310.2006.01125.x.

29. Kaye, J. N. (2001). Symbolic Olfactory Display. MIT. Cambridge, MA, http://hdl.handle.net/1721.1/16788.

30. Batch, A., B. Patnaik, M. Akazue, & N. Elmqvist. (2020). "Scents and Sensibility: Evaluating Information Olfactation." Proceedings of the 2020 CHI Conference on Human Factors in Computing Systems, 1–14, https://doi.org/10.1145/3313831.3376733; Patnaik, B., A. Batch, & N. Elmqvist. (2019). "Information Olfactation: Harnessing Scent to Convey Data." *IEEE Transactions on Visualization and Computer Graphics*, 25(1), 726–736, https://doi.org/10.1109/TVCG.2018.2865237.

31. Batch, A., Patnaik, B., M. Akazue, & N. Elmqvist. (2020). "Scents and Sensibility: Evaluating Information Olfactation." Proceedings of the 2020 CHI Conference on Human Factors in Computing Systems, 1–14, https://doi.org/10.1145/3313831.3376733.

32. Dmitrenko, D., E. Maggioni, & M. Obrist. (2017). "OSpace: Towards a Systematic Exploration of Olfactory Interaction Spaces." Proceedings of the 2017 ACM International Conference on Interactive Surfaces and Spaces, 171–180, https://doi.org/10.1145/3132272.3134121.

33. Obrist, M., A. N. Tuch, & K. Hornbaek. (2014). "Opportunities for Odor: Experiences with Smell and Implications for Technology." Proceedings of the SIGCHI Conference on Human Factors in Computing Systems, 2843–2852, https://doi.org/10.1145/2556288.2557008.

34. Dobbelstein, D., Herrdum, S., & Rukzio, E. (2017). "Inscent: A Wearable Olfactory Display as an Amplification For Mobile Notifications." Proceedings of the 2017 ACM International Symposium on Wearable Computers, 130–137, https://doi.org/10.1145/3123021.3123035.

35. Matchar, E. (2019, February 8). "The Pharmacist Who Launched America's Modern Candy Industry | Innovation| Smithsonian Magazine." *Smithsonian Magazine*, https://www.smithsonianmag.com/innovation/pharmacist-who-launched-americas-modern-candy-industry-180971354/.

36. Kirsner, S. (2018, May 4). "A Factory in Cambridge Makes 14 Million Junior Mints a Day. Why Is No One Allowed Inside?—The Boston Globe." *Boston Globe*, https://www.bostonglobe.com/business/2018/05/04/junior-mints-and-more-this-factory-makes-pieces-candy-day/6PnudsUCsxbqsi19VvKuZJ/story.html.

37. Wijnsma, L. & F. Tan. (2017). "Smell of Data," https://smellofdata.com/.

38. Alice, O. & K. McLean. (n.d.). "100,000 Is a Lot of People." Retrieved July 31, 2023, from http://sensorymaps.blogspot.com/2013/04/100000-is-lot-of-people.html.

39. McLean, K. (2020). *Temporalities of the Smellscape: Creative Mapping as Visual Representation*, 217–246. Springer. Berlin, Germany. http://doi.org/10.1007/978-3-658-30956-5.

REFERENCES   235

40. Mackay, R. (2021, December 14). "The Smell of Success: How Scent Became the Must-Have Interpretative Tool." *Blooloop*, https://blooloop.com/museum/in-depth/museum-scents/.
41. Bembibre, C. & M. Strlič. (2017). "Smell of Heritage: A Framework for the Identification, Analysis and Archival of Historic Odours." *Heritage Science*, 5(1), 2, https://doi.org/10.1186/s40494-016-0114-1.
42. Verbeek, Caro. (2021, January 15). "What Can The Scents Of The Past Tell Us About Our History?." In *TED Radio Hour*. NPR, https://www.npr.org/2021/01/15/956933669/caro-verbeek-what-can-the-scents-of-the-past-tell-us-about-our-history
43. (2015). *National Museum and Jetaime Perfumery Interprets 700 Years Of Singapura Through Smells*. Jetaime Perfumery Pte Ltd. Singapore. Retrieved July 31, 2023, from https://www.newswire.com/news/national-museum-and-jetaime-perfumery-interprets-700-years-of.
44. Belkin, K., R. Martin, S. E. Kemp, & A. N. Gilbert. (1997). "Auditory Pitch as a Perceptual Analogue to Odor Quality." *Psychological Science*, 8(4), 340–342, https://doi.org/10.1111/j.1467-9280.1997.tb00450.x.
45. Hanson-Vaux, G., A.-S. Crisinel, & C. Spence. (2013). "Smelling Shapes: Crossmodal Correspondences between Odors and Shapes." *Chemical Senses*, 38(2), 161–166, https://doi.org/10.1093/chemse/bjs087
46. Gayler, T., C. Sas, & V. Kalnikaitė. (2022). "Exploring the Design Space for Human-Food-Technology Interaction: An Approach from the Lens of Eating Experiences." *ACM Transactions on Computer-Human Interaction*, 29(2), 1–52, https://doi.org/10.1145/3484439.
47. Wang, Y., X. Ma, Q. Luo, & H. Qu. (2016). "Data Edibilization: Representing Data with Food." Proceedings of the 2016 CHI Conference Extended Abstracts on Human Factors in Computing Systems, 409–422, https://doi.org/10.1145/2851581.2892570.
48. Gayler, T., C. Sas, & V. Kalnikaite. (2023). "'It Took me Back 25 Years in One Bound': Self-Generated Flavor-Based Cues for Self-Defining Memories in Later Life." *Human–Computer Interaction*, 38(5–6), 417–458, https://doi.org/10.1080/07370024.2022.2107518.
49. Krenn, V. & V. Mihaylova. (n.d.). "Taste of Data." [Tumblr]. Prazlab. Retrieved August 9, 2023, from https://taste-of-data.tumblr.com/.
50. Sayej, N. (2014, August 4). "Austrian Designers Turn Corruption into Edible Infographics." *Vice*, https://www.vice.com/en/article/z4gdw3/austrian-designers-turn-corruption-into-edible-infographics
51. Dolejšová, M. (2015). "A Taste of Big Data on the Global Dinner Table." *Journal for Artistic Research*, 9, https://www.researchcatalogue.net/view/57801/57823/.
52. International Secretariat. (2012). *Corruption Perceptions Index 2012*. Transparency International. Berlin, Germany.
53. CBS Mornings [@CBSMornings], & Dokoupil, T. (2020, January 31). "WEALTH INEQUALITY: The Richest 1% Controls more Wealth now than at Any Time in more than 50 Years. But what Does Wealth Inequality Really Look Like? @TonyDokoupil turned America's Economic Pie into a Real One and Asked People a Simple Question: Who Gets What?" Https://t.co/scGPKcHbie [Tweet]. Twitter, https://twitter.com/CBSMornings/status/1223224310413905921.
54. McDowell, C. (2015, June 11). "Big Democracy: All In. Interaction Institute for Social Change," https://interactioninstitute.org/all-in/.
55. Ancel, D. & S. Girel. (2014). "Art and the Public Space." In *Kunst und Öffentlichkeit*, D. Danko, O. Moeschler, & F. Schumacher (eds.), 83–93, Springer VS. Wiesbaden, Germany. https://doi.org/10.1007/978-3-658-01834-4_5.
56. Happy City Denver. (2018). "Art for the People Experiments Report." Denver Theatre District, https://issuu.com/denvertheatredistrict/docs/happy_city_report_-_denver.
57. Webb, D. (2014). "Placemaking and Social Equity: Expanding the Framework of Creative Placemaking." *Artivate*, 3(1), 35–48, https://doi.org/10.1353/artv.2014.0000.
58. Courage, C. (2020). "The Art of Placemaking: A Typology of Art Practices in Placemaking." In *The Routledge Handbook of Place*. Routledge. London, UK.
59. Lindgaard, K. & H. Wesselius. (2017). "Once More, with Feeling: Design Thinking and Embodied Cognition." *She Ji: The Journal of Design, Economics, and Innovation*, 3(2), 83–92, https://doi.org/10.1016/j.sheji.2017.05.004.

Conclusion

1. Their name is a tongue-in-cheek reference to the community's contentious relationship with the Los Angeles Police Department—the "other" LAPD. They've deliberately appropriated the acronym and put it in service of their causes.

236 REFERENCES

2. Alpert-Reyes, E. (2015, January 9). "Activists Fear that Big-Project Zoning Change would Ignore L.A.'s Poor." *Los Angeles Times,* https://www.latimes.com/local/cityhall/la-me-big-development-20150109-story.html.

3. It isn't as well-known globally; while working with colleagues in the Brazilian city of Belo Horizonte on a data mural I stumbled across a mini-golf course on the top floor of a shopping mall and amusingly had to explain the background to my local friends. We stopped to play the first round ever.

4. Thomas, R. M. Jr. (1996, April 18). "Don Clayton, 70, Driven Man Who Putted His Way to Riches." *The New York Times,* https://www.nytimes.com/1996/04/18/us/don-clayton-70-driven-man-who-putted-his-way-to-riches.html.

5. PUTTING GREEN Mini Golf|Brooklyn, NY. (n.d.). Retrieved October 30, 2023, from https://www.puttinggreenbk.org/.

6. Fotopoulou, A., H. Barratt, & E. Marandet. (2021). "A Data-Based Participatory Approach for Health Equity and Digital Inclusion: Prioritizing Stakeholders." *Health Promotion International,* daab166, https://doi.org/10.1093/heapro/daab166.

7. McLeroy, K. R., D. Bibeau, A. Steckler, & K. Glanz. (1988). "An Ecological Perspective on Health Promotion Programs." *Health Education Quarterly,* 15(4), 351–377, https://doi.org/10.1177/109019818801500401.

8. *Edith Ackermann's Pedagogical Perspective on Tinkering & Making.* (2014, August 23). The Exploratorium, https://vimeo.com/104178407.

9. Harel, I. & S. Papert. (eds.). (1991). "Situating Constructionism." In *Constructionism.* Ablex Publishing. New York.

10. Fuller, R. B. (2008). *Operating Manual for Spaceship Earth* (first edn). Lars Muller. Zürich, Switzerland.

11. Lan, X., Y. Wu., & N. Cao. (2023). "Affective Visualization Design: Leveraging the Emotional Impact of Data." IEEE Transactions on Visualization and Computer Graphics, 1–11, https://doi.org/10.1109/TVCG.2023.3327385.

12. Campbell, S. & D. Offenhuber. (2019). "Feeling Numbers: The Emotional Impact of Proximity Techniques in Visualization." *Information Design Journal,* 25(1), 71–86, https://doi.org/10.1075/idj.25.1.06cam; Peck, E. M., S. E. Ayuso, & O. El-Etr. (2019). "Data is Personal: Attitudes and Perceptions of Data Visualization in Rural Pennsylvania." Proceedings of the 2019 CHI Conference on Human Factors in Computing Systems—CHI '19, 1–12, https://doi.org/10.1145/3290605.3300474.

13. Dick, M. (2020). *The Infographic.* MIT Press. Cambridge, MA.

14. Veysey, I. (2016). "A Statistical Campaign: Florence Nightingale and Harriet Martineau's England and her Soldiers." *Science Museum Group Journal,* Spring, https://dx.doi.org/10.15180/160504/001.

15. Du Bois, W. E. B., W. Battle-Baptiste, & B. Rusert. (2018). *W.E.B Du Bois's Data Portraits: Visualizing Black America* (first edn). The W.E.B. Du Bois Center At the University of Massachusetts Amherst; Princeton Architectural Press.

16. Charlton, J. I. (2000). *Nothing About Us Without Us: Disability Oppression and Empowerment.* University of California Press. Berkeley, CA.

17. Thorp, J. (2021). *Living in Data: A Citizen's Guide to a Better Information Future.* MCD. New York, New York.

18. Vasdev, S. (2020). *Embracing Creative Arts to Amplify Data Use.* Tanzania Bora Initiative. Dar es Salaam, Tanzania.

19. Tanzania Bora (Director). (2018, November 3). "Data za kitaa Episode 1 with English Subtitles," https://www.youtube.com/watch?v=5lVTjEV831s.

20. Tanzania Bora (Director). (2018, November 5). "Narudi NyumbanI Official Video: KINASA," https://www.youtube.com/watch?v=5B7ru4V_mpQ.

21. Ayanna Pressley [@AyannaPressley]. (2018, June 30). (1/11) When I say..."The people closest to the pain, should be the closest to the power, driving & informing the policymaking ..." THIS is what I mean. (A) When I wanted to break cycles of #poverty by strengthening pathways to #graduation, preventing #dropout #MA7 [Tweet]. Twitter, https://twitter.com/AyannaPressley/status/1013184081696346113.

Index

For the benefit of digital users, indexed terms that span two pages (e.g., 52–53) may, on occasion, appear on only one of those pages.

Figures are indicated by an italic *f* following the page number. Note information is indicated by n and note number following the page number.

accessibility of data. *see* data accessibility
actuation, 116
agenda setting, 36
Agloe, New York, 49–50
air quality, taste reflecting, 188–189, 189*f*
Alice, Olivia, 185
alternative imaginaries, 46
artificial intelligence, 7–8, 148
arts-based approaches
  data murals as. *see* data murals
  data protests as, 16–23, 19*f*. *see also* data
    protests
  data sculptures as. *see* data sculptures
  data theatre as. *see* data theatre
  multi-sensory. *see* multi-sensory data
    experiences
  popular data inclusion of, 10. *see also* popular
    data
  questions and answers via, 10
Association for Computing Machinery, 62
astronomy, sonification in, 174, 178
audification, 173–174. *see also* sonification
auditory icons, 173
Aushabuc, Aikon, 157
autographic visualization, 81

Back 9, The, 200–203, 202*f*, 203*f*, 204*f*, 205–210
Bacon, Ben, 24
Bankslave, 80–81, 81*f*
barriers to participation
  data literacy and, 108–109
  data sculptures breaking down, 115
  democracy hindered by, 44
  layers of reading and, 136
  power raising, 11
BBC (British Broadcasting
  Corporation), 172–175
Before I Die murals, 97–98
Belliveau, George, 152
Bembibre, Cecilia, 185–186

Bertin, Jacques, 141–142, 142*f*, 144–145
Bhoomi system, 93
Big Data
  abolition of, 46
  alternatives to focus on, 7–8
  centralization of, 29
  definition of, 28
  extraction of, 28
  fields of origin underlying, 28–29
  opaqueness of, 28
  solutionism focus on, 10
  technological complexity of, 28–29
  zine coverage of, 148
Black communities
  data justice for, 68–70, 101–102
  data literacy in, 101–102
  data protests in, 109–111
  data sculptures of/for, 109–111, 117,
    118–119, 127–128
  data visualization omitting, 140–141
  data visualization representing, 211
  education/literacy denial for, 100–101
  feminism by women of color in, 224 n.43
  food insecurity in, 122
  journalists reporting on, 33–38
  racist data on. *see* racist data
  redlining in, 85–86, 86*f*
Black Lives Matter movement, 109–110
Black Youth Project 100 (BYP100), 109–111,
  117, 118–119
Boal, Augusto, 160–163
bodily data sorting, 157–158
body syntonicity, 155–156, 231 n.61
Borneman, Elizabeth, 140–141
Bostock, Mike, 37
Boston Public Library, 31–32
Brea, Amanda, 151–152
Brecht, Bertolt, 161
British Broadcasting Corporation
  (BBC), 172–175

238 INDEX

Brooks slave ship, 125–127, 126*f*
Brouwers, Henriëtte, 201
Brukilacchio, Lisa, 1
Bus Regulation: The Musical, 162
BYP100 (Black Youth Project 100), 109–111, 117, 118–119
C4HW (Collaborate for Healthy Weight Coalition) initiative, 1–5, 2*f*, 78*f*, 77–79

Cairo, Alberto, 139, 141
candy, smell of, 184
CARE principles, 64
Carr, Jonathan, 152
Carter, Ennis, 52–54
cartoons, 150
cave paintings, 23*f*, 23–24
centering impact. *see* impact of data
Chalabi, Mona, 144
Chan, Paul, 162
Chang, Candy, 97–98
charts and graphs
  as barriers to participation, 44
  data literacy with, 32
  data murals and graffiti depicting, 74*f*, 74
  data protests *vs.*, 19–20
  data sculptures depicting, 110–111, 118–120, 119*f*, 128–129
  effectiveness of, 144–147, 145*f*, 212–213, 229 n.24
  haptic navigational, 111–113, 112*f*
  historical, 211
  layers of reading beyond, 136, 212–213
  limitations of, 1, 110, 124–125, 134–136, 139, 210–213
  living histograms depicting, 157, 158*f*, 162, 170–171
  multi-sensory data experiences *vs.*, 168–169
  participatory data graphs as, 57–58
  taste data *vs.*, 190, 192
  3D, 118–120, 119*f*, 224 n.47
  2D, 141–144, 211, 212–213
Chicano Mural Movement, 51–52
City Life/Vida Urbana, 85–86, 86*f*
Ciuccarelli, Paolo, 178–179
Civic Switchboard Project, 32–33
civil society organizations (CSOs)
  collective action via, 47–49
  data and, 49
  data justice goals of, 70–71
  definition of, 48
  feminist, 71–72
  impact of data from, 21–22, 29, 47–49
  multi-sensory data experiences with, 179–180, 197

origins of, 48, 222 n.75
popular data application by, 14–15, 199–210
zine use by, 147–148, 150
Clark, Duncan, 37
Clarkson, Thomas, 125–127
Cleveland, Bill, 27–28
climate change
  mini-golf course on, 201
  museum exhibits addressing, 38–40, 39*f*
  music on, 176–177
ClimateMusic Project, 176–177
clothing sizes, women's, 13, 88, 89–91, 90*f*, 212
cocktail party effect, 174
Collaborate for Healthy Weight Coalition (C4HW) initiative, 1–5, 2*f*, 77–79, 78*f*
collective action, 47–49
colonialism. *see* data colonialism
comics, 150
commercials, data embodiment in, 158–160
community, defined, 20
community data theatre, 163–165
complexity, layers of reading to explore. *see* layers of reading
computational literacy, 99
Conference on Fairness, Accountability, and Transparency (FAccT), 62
Connected by Data project, 63–64
constructionism, 127, 155–156, 208
constructivism, 155–156
Cooper, Marshal, 85, 86*f*
copyright traps, 49
counterdata production, 17
COVID-19 Model Mountain, 119*f*, 119–120
COVID-19 pandemic
  data murals reflecting situation in, 82–84, 83*f*, 84*f*
  data sculptures on, 118–125, 119*f*, 121*f*, 124*f*
  food insecurity in, 121–125, 123*f*, 124*f*
critical data literacy, 64–67, 65*f*, 79
critical theories of data use, 61–73
CSOs. *see* civil society organizations
Cukier, Kenneth, 91
culinary design, 191–193
culinary variables, 191
cultural context
  data fashion shows reflecting, 13–14, 131–132, 134–135, 214
  data murals reflecting, 2–3, 79
  data theatre exploring, 151–152
  data visualization norms and, 139–140
  LAPD functioning in, 207
  museums reflecting, 38–43, 39*f*
  taste and, 187–193
Cuomo, Andrew, 119*f*, 119–120, 227 n.61

cutlery, data sculpture of, 121–125, 124*f*
cymatics, 167–168

d3.js, 37
D4BL (Data for Black Lives) conference, 68–70,
    69*f*, 127–128
Dana, John Cotton, 41
dance, 157–158
data
    accessibility of. *see* data accessibility
    arts-based approaches to. *see* arts-based
        approaches
    defining, 25–27, 140
    density of, 139
    fields of origin for, 27–29
    gender and. *see* data feminism; women
    for good. *see* data for good
    history of, 23–29, 25*f*, 95*f*, 94, 213–214
    impact of. *see* impact of data
    justice and injustice with. *see* data justice;
        injustice; power
    literacy with. *see* data literacy
    open data movement on, 8, 45–47, 148
    popular data methodology for. *see* popular
        data
    storytelling with. *see* data storytelling
Data2X, 62
Data4Change, 62, 82–83, 179–180
data accessibility
    data murals improving, 83–84
    libraries increasing, 32–33
    multi-sensory data experiences
        broadening, 179–180, 193, 194–198
    museum goals of, 40–41
    open data movement increasing, 8, 45–47,
        148
    power and injustice connections to, 9, 11,
        213–214
    toolbox for, 1
DataBasic.io project, 6
data colonialism, 60, 63, 64, 132–133. *see also*
    data decolonization
Data Cuisine workshops, 191–192
Data Culture Project, 6, 9, 224 n.51
data decolonization, 61, 63–64. *see also* data
    colonialism
data density, 139
data fashion shows
    culture reflected in, 13–14, 131–132,
        134–135, 214
    layers of reading via, 13–14, 131–136, 132*f*,
        133*f*, 212–213
    participation via, 134–135
data feminism

definition of, 72
feminism, defined for, 71–72, 224 n.43
feminist standpoint theory and, 139
guiding principles for, 72
participation and, 13, 71–73, 212
datafication, creating windows, 91–93, 212
Data for Black Lives (D4BL) conference, 68–70,
    69*f*, 127–128
data for good
    critiques of movement for, 62–64
    multi-sensory data experiences and, 197
    participation and, 13, 61–64, 212
    popular data to support, 10–11. *see also*
        popular data
    pro-social spaces' focus on, 20
data graffiti, 74, 74*f*, 80–82, 81*f*, 82*f*, 84
data-ink ratio, 139
data justice
    data feminism goals of, 71–72
    data literacy and, 99, 101–102
    data murals as tool for, 54, 69–70
    democracy and, 71
    impact of data for, 45
    participation and, 13, 67–71, 212
    social justice and, 70–71
Data Justice Lab, 69–70
Data Khanga Fashion Show, 131–136, 132*f*,
    133*f*. *see also khangas*
DataKind, 62
data literacy
    beneficiaries of, 108–109
    concept of, 98–109
    critical, 64–67, 65*f*, 79
    data as window and, 13, 91, 93, 98
    data murals transcending, 3–4, 54, 79
    data sculptures as tool for, 109, 114–115
    definition of, 99, 106–108
    education for, 64–67, 65*f*, 99, 101–102, 109
    embodied learning and, 157–158
    libraries promoting, 32–33, 109
    math skills and, 102–106, 105*f*, 108
    mirrors and, 98–109, 212
    participation and, 13, 64–67, 65*f*, 79, 212
    power from, 100–102
    social transformation as goal of, 64
data murals
    artistic *vs.* empirical, 176
    data justice with, 54, 69–70
    definition of, 1–2, 73
    design goals for, 55–56
    global use of, 82–85, 83*f*, 84*f*
    graffiti and, 74, 74*f*, 80–82, 81*f*, 82*f*, 84
    history of muralism and, 51–52
    how to make, 76–79

240 INDEX

data murals (*Continued*)
  as mirrors, 97–98
  multi-sensory data experiences and, 197
  participation through, 13, 51–56, 53*f*, 73–85,
    74*f*, 75*f*, 76*f*, 78*f*, 81*f*, 83*f*, 84*f*, 212
  unifying nature of, 3, 4*f*
Data.org, 62
data protests
  data fashion shows for, 13–14, 131–133,
    212–213
  data sculptures for, 109–111, 125–127, 126*f*
  impact of data in, 12–13, 19*f*, 16–23, 211–212
data science, 27–28, 92*f*, 91
Data Science and Literacy Act, 99
Data Science for Social Good, 62. *see also* data
  for good
data sculptures
  as activism, 125–127, 126*f*
  artistic, 120–125, 121*f*, 176
  data literacy via, 109, 114–115
  data protests with, 109–111, 125–127, 126*f*
  definition of, 116
  design and approaches to, 116–118, 121
  empirical, 176
  haptic visualization with, 111–114, 112*f*,
    112*f*, 113*f*
  intent and motivation for, 117, 127
  interaction with, 127–129
  layers of reading with, 128–129
  as learning tools, 114–116
  libraries including, 30–31
  as mirrors, 13, 91, 109–130, 212
  multi-sensory data experiences and, 186, 197
  participation via, 114–116, 127–129
  popularity and number of, 129
  real world examples of, 118–129, 119*f*
  as spaces of possibilities, 116–118
  techniques to create, 116
  3D charts as, 118–120, 119*f*
  virtual, 116, 129–130
Data & Society, 69–70
data sonification. *see* sonification
data storytelling
  arts-based approaches to. *see* arts-based
    approaches
  definition of data and, 26–27
  family photo analogy to, 8–9
  layers of reading for. *see* layers of reading
  narrative of, 143–144
  participation in. *see* participation
  popular data as method for, 10. *see also*
    popular data
  power shaping, 11
  toolbox for, 1

workflow or model for, 55*f*, 55–56
  workshops for, 5
data strings installation, 128*f*
data theatre
  commercials with, 158–160
  community, 163–165
  as embodiment and embodied
    learning, 154–158, 161, 163, 165
  exploration and development of, 150–165,
    153*f*
  LAPD popular data including, 14–15, 204*f*,
    199–200, 202–203, 207–208
  layers of reading via, 13–14, 150–165,
    212–213
  multi-sensory data experiences and, 197
  participatory theatre and, 160–163, 207–208
Data Therapy project, 5
data visualization
  arts-based approaches to. *see* arts-based
    approaches
  charts and graphs for. *see* charts and graphs
  data feminism on, 72
  data literacy and, 107
  definition of, 139–141
  government use of, 44
  haptic, 111–114, 112*f*, 112*f*, 113*f*
  interactive. *see* interactive exhibits
  journalism and, 37
  museums and, 42, 221 n.65. *see also*
    interactive exhibits
  norms of, problematic, 137–150
  2D, 141–144, 211, 212–213
Data Za kitaa/Street Data, 214
Data Zetu, 60–61, 63, 84–85, 133–134, 229 n.7
Dear Data (Lupi and Posavec), 42
decolonization. *see* data decolonization
democracy
  data and, 45–46
  data justice and, 71
  impact of data for healthy, 21, 29, 43–46
  journalism supporting, 36
  open data movement and, 45–46
  participation as foundational to, 43–46, 196
  popular data improving, 213–214
  power and, 44–45
demonstrations. *see* data protests
Descartes, René, 154–155
DesPortes, Kayla, 157–158
Detroit Community Technology Project, 69–70
Detroit Digital Justice Coalition, 49
dialectical theatre, 161
Diaz Merced, Wanda, 174, 178
digital literacy, 99

digital production of data sculptures, 116, 129–130
digital sonification, 177–179, 178*f*
D'Ignazio, Catherine, 6, 17, 72–73, 104, 109
Dokkl library, 29–31
Dokoupil, Tony, 194–196, 195*f*
Domestic Data Streamers, 127–128, 128*f*
Downey, Allen, 102–103
Duarte, Jose, 74, 224 n.47
DuBois, W. E. B., 211

earcons, 173
edibilization, 189–190, 192–194, 197, 213. *see also* taste
education. *see also* workshops
    data literacy, 64–67, 65*f*, 99, 101–102, 109
    data science, 91, 92*f*
    data sculptures for, 114–116
    data theatre as embodied, 154–158
    data visualization, 137–138
    denial of, 100–101
    impact of data to increase, 12–13
    informal, 22–23
    math, 102–106
    multiple ways of knowing in, 169–171
    museums providing, 41–42
    participatory data graphs on, 57
    popular, 9–10, 101, 150, 160
    textbooks for, 47–49
    zines for, 150
efficacy
    data murals building, 54
    impact of data to increase, 12–13, 29, 45
    participation increasing, 54, 61
    popular data as tool for, 10–11, 213–214
embodied cognition, 156–157
embodiment
    commercials as tool for, 158–160
    data theatre as, 154–158, 161, 163, 165. *see also* data theatre
    definition of, 111
Embody a Dataset activity, 163
empathy
    data theatre and, 155–156, 163–165
    embodied learning increasing, 155
    multi-sensory data experiences increasing, 179–180, 187
employment
    barriers to participation and, 11
    data murals on, 13, 51–55, 53*f*, 80
    history of data in, 24
    punitive data use for, 21–22, 95*f*, 93–94
empowerment. *see also* power
    data accessibility for, 1

    data decolonization for, 63–64
    data literacy for, 99–102, 107, 108
    data murals for, 54, 73–76
    data theatre for, 161–162
    impact of data to increase, 12–13, 29, 45
    mirrors for, 95
    multi-sensory data experiences for, 197–198
    popular data as tool for, 10–11, 205–206, 210–215
empty chairs, as data sculpture, 120–121, 121*f*
energy
    chart effectiveness for information on, 146–147
    data murals on, 74–75, 75*f*
    mirrors of data on, 96*f*, 96–97
    museum exhibits on, 39–40
engagement. *see also* participation
    data accessibility for, 1
    data literacy for, 107–108
    data theatre inviting, 151, 162, 164–165
    layers of reading and, 136, 165
    mirrors for, 95
    multi-sensory data experiences offering, 171, 176–177, 186–188, 193, 194–198
    murals/data murals for, 51–52, 54, 73–74
    popular data as tool for, 10–12
environment. *see also* location
    climate change effects on. *see* climate change
    data sculptures specific to, 116
    multi-sensory data experiences on, 166–169, 167*f*, 176–177, 188–189, 189*f*
epic theatre, 161
epistemological pluralism, 171
epistemology, 170
Errazuriz, Sebastian, 75, 76*f*
ethics, 61–64, 67, 72
eugenics, 88–89
Excel, 141–143, 229 n.24
experiential knowledge, 170–171, 177, 180, 195–196, 210
extended epistemology, 170

FAccT (Conference on Fairness, Accountability, and Transparency), 62
fashion shows. *see* data fashion shows
feminism. *see* data feminism
feminist standpoint theory, 139
fields of origin for data, 27–29
fifth estate, 36
Financial Times
    sonification by, 179
    Visual Vocabulary Guide, 143*f*
Flourish, 37
focus on participation. *see* participation

242 INDEX

food
  insecurity of. *see* food insecurity
  multi-sensory engagement with, 194–196, 195*f*
  smell of, 184–186
  taste of. *see* taste
food insecurity
  data sculpture on, 121–125, 123*f*, 124*f*
  data theatre on, 164
Forum Theatre, 160–161
4th wall, breaking through, 160–161
fourth public estate, 36. *see also* journalism
Fragapane, Federica, 162
framing, by journalism, 36
Freire, Paulo, 9–10, 66–67, 99, 101, 150, 160
functionalist-idealist discourse, 140

Geere, Duncan, 179
General Drafting, 49–50
gestures, 156–157
Gilbert, Daniel, 159
Glass Room, The, 42
Go Boston 2030, 58–61, 59*f*
Golden, Jane, 52
Goldin-Meadow, Susan, 156–157
good, data for. *see* data for good
Google, 109, 177–178, 229 n.24
Got Data workshops, 5
government. *see also* democracy
  civil society organizations and, 47–49
  datafication by, 92–93
  data literacy support by, 99
  data protests influencing, 16–17
  data sculptures by, 118
  fields of origin for data for, 27
  history of data of, 24
  impact of data from, 21–22, 29, 43–46
  open data movement for transparency of, 8, 45–47
  urban planning by. *see* urban planning
graffiti. *see* data graffiti
graphs. *see* charts and graphs
Gray, R. E., 152
Greenland tactile maps, 113*f*, 113
Gurian, Elaine Heumann, 41
G-Watch, 47

handcrafting, of data sculptures, 116
Hansen, Mark, 162–163
haptic memory, 111
haptic visualization, 111–114, 112*f*, 112*f*, 113*f*
Harris, Rich, 37
Harrison, Ellie, 162
Head Start, 2–3, 2*f*, 77–78, 224 n.50

hearing data. *see* sonification
Hear the Blind Spot project, 179–180
Hermann, Thomas, 173
Hinson, Jesse, 151–152
Holm, Gustav, 113
Holmes, Nigel, 144
housing rights, 14–15, 199–210. *see also* redlining
How to House 7,000 People in Skid Row, 200, 203–210, 205*f*, 206*f*
How We Fish mural, 13, 51, 52–56, 53*f*, 73, 78, 212
human immunodeficiency virus (HIV) transmission, 131
hyperosmia, 180

ICAD (International Community for Auditory Displays), 174–175
IISC (Interaction Institute for Social Change), 58–60
IKEA, 139–140
Image Theatre, 160–161
impact of data
  centering of, 12–13, 16, 206–207, 211–213
  from civil society organizations, 21–22, 29, 47–49
  data protests highlighting, 12–13, 16–23, 19*f*, 211–212
  defining data and, 25–27
  fields of origin for data and, 27–29
  from government, 21–22, 29, 43–46
  history of data and, 23–29, 25*f*
  from journalism, 21–22, 29, 33–38
  in LAPD projects, 206–207, 207*f*
  from libraries, 21–22, 29–33
  from museums, 21, 29, 38–43, 39*f*
  owning personal, 49–50
  power as, 21–22, 28, 50
  pro-social spaces' lessons on, 12–13, 20–23, 29–49
inclusion. *see also* participation
  civil society organizations encouraging, 48
  data visualization norms without, 140–141
  murals for, 51–52
  museum as space for, 41
  popular data as tool for, 205–206, 213–214
indexical data sculptures, 117, 128–129
Indigenous communities
  data decolonization goals of, 64
  data justice for, 68
  haptic visualization in, 111–113, 112*f*, 112*f*, 113*f*
  impact of data for, 50
  murals for inclusion of, 51–52

INDEX 243

infographics, 37, 42, 126, 157, 211, 214
informational inferential reasoning, 155–156
information literacy, 99
information receptivity, 146–147
injustice
  data accessibility connections to, 9, 11
  data colonialism creating, 63
  data graffiti addressing, 80
  data justice to address. *see* data justice
  data literacy uncovering, 67
  data murals addressing, 80, 82–83
  data protests to address, 109–110
  impact of data for, 21–22
inScent project, 183–184
Interaction Institute for Social Change
    (IISC), 58–60
interactive exhibits
  LAPD popular data including, 14–15, 199,
    203–210, 205f, 206f
  layers of reading via, 13–14, 40, 135–136
  as mirrors, 96–98, 96f, 97f
  museums creating, 13–14, 39f, 38–42, 221
    n.65
International Community for Auditory Displays
    (ICAD), 174–175
Inuit maps, 113f, 113
Investigative Reporters and Editors Inc., 109
Invisible Theatre, 160–163
ionic data sculptures, 117
Isotype (Neurath and Neurath), 42, 43f

Jaschko, Susanne, 191, 193
Johnson, Léa, 191f, 191–192
Joiner, Brian, 157, 158f, 162, 170–171
Jorvik Viking Centre, 185–186
journalism
  data literacy for, 109
  data or digital, 36–38
  data sculptures in, 129–130
  data theatre reflecting, 160–161
  impactful stories by, 33–38
  impact of data from, 21–22, 29, 33–38
  investigative, 35, 37
  layers of reading in, 135–136
  multi-sensory data experiences in, 194–197,
    195f, 214
  participatory data graphs in, 57–58
  societal role of, 35–37
  sonification in, 172–173, 175, 179
  taste-based data in, 188–189
  zines for, 147–150, 149f
Just Data Cube, 68–69, 69f, 127–128
justice. *see* data justice; injustice

Kamp, John, 114
Kaufmann, Dorsey, 167–169
Kaye, Joseph, 182
Ketchum, Alex, 148
khangas, 132–133, 133f, 228 n.4, 134–135,
    139–141, 212–214. *see also* Data Khanga
    Fashion Show
kinesthetic intelligence, 155
Klein, L. F., 72
Knight Center, 109
Krenn, Veronica, 192–193
Kunit, 113
Kunze, Jane, 30

Lacy, Suzanne, 162
land ownership, data on, 93
language
  as barrier to participation, 11
  data theatre overcoming barriers of, 160–161
  definition of data tied to, 26–27
  gestures *vs.* spoken, 156–157
  history of data and written, 24
LAPD. *see* Los Angeles Poverty Department
layers of reading
  broadening approaches to enhance, 147–150
  building to complexity with, 165
  chart effectiveness and, 144–147, 145f, 229
    n.24, 212–213
  creating, 135–136, 165
  data fashion shows for, 13–14, 131–136, 132f,
    133f, 212–213
  data murals requiring, 79
  data sculptures offering, 128–129
  data theatre for, 13–14, 150–165, 212–213
  data visualization definition and, 139–141
  data visualization norms and, 137–150
  interactive museum exhibits for, 13–14, 40,
    135–136
  LAPD projects creating, 209–210
  participation and, 136
  popular data to create, 13–14, 131, 209–210,
    212–213
  taste data creating, 190–191, 193
  2D representations *vs.*, 141–144, 212–213
Lenzi, Sara, 178–179
Leonard, Alison, 103
Levin, Golan, 74f, 74
Levi-Strauss, C., 100–101, 225 n.19
Li, Chrystal, 148
Liberatory Computing project (MIT), 101–102
libraries. *see also* pro-social spaces
  data and, 32–33
  data literacy and, 32–33, 109
  history of, 31–32

244   INDEX

libraries. *see also* pro-social spaces (*Continued*)
  impact of data from, 21–22, 29–33
  multi-sensory experiences with, 197
  public knowledge curation by, 29–33
  zine use by, 147–149
Life Under Curfew data murals, 82–84, 83*f*, 84*f*
LivableStreets Alliance data theatre, 152–154, 153*f*
living histograms, 157, 158*f*, 162, 170–171
Local Lotto project, 64–67, 65*f*
location
  data murals specific to, 82
  data sculptures specific to, 116
  redlining focus on, 86
Loomis, Wendy, 176
Los Angeles Poverty Department (LAPD)
  context for work of, 199–200
  data theatre by, 14–15, 204*f*, 199–200, 202–203, 207–208
  impact of data from, 206–207, 207*f*
  interactive exhibits by, 14–15, 199, 203–210, 205*f*, 206*f*
  layers of reading in projects by, 209–210
  mini-golf course by, 14–15, 199, 200–203, 202*f*, 203*f*, 204*f*, 205–210
  mirrors built by, 208–209
  multi-sensory data experiences by, 210
  name of, 235 n.1
  participation encouraged by, 207–208
  popular data application by, 14–15, 199–210
Loud Numbers podcast, 179
Lowenfeld, Viktor, 155
Lowery, Wesley, 38

Mackinlay, Jock, 144–145, 145*f*, 229 n.24
Mages, Michael Arnold, 152
Make a (Data) Scene activity, 163–164
Malpede, John, 200–203
Map of the Future, The, 38–41, 39*f*
Marco, Danford, 134
Marklein, Alison, 176
marks, for visual encoding, 141–142
Marshall Islands navigational charts, 111–113, 112*f*
Mason-Brown, Lucas, 68
math skills, 102–106, 105*f*, 108
Mayer-Schönberger, Viktor, 91
McDowell, Cesar, 196
McLean, Kate, 185
media, news. *see* journalism
Memorandum on Transparency and Open Government (2009, US), 46
Mexican muralism, 51–53
Mihaylova, Veselea, 192–193

Milan smell map, 185*f*, 185
Milner, Yeshimabeit, 68
mind-body duality, 154–155
mini-golf courses, 14–15, 199, 200–203, 202*f*, 203*f*, 204*f*, 205–210, 236 n.3
mini-publics, 45
mirrors
  benefits of data as, 95
  data literacy and, 98–109, 212
  data murals as, 97–98
  data protests as, 18
  data sculptures as, 13, 91, 109–130, 212
  haptic visualization and, 111–114, 112*f*, 112*f*, 113*f*
  interactive exhibits as, 96–98, 96*f*, 97*f*
  LAPD projects as, 208–209
  popular data to build, 13, 88, 208–209, 212
  racist data not as, 13, 88–98, 212
  windows *vs.*, 13, 88–95, 98, 212
Mockus, Antanas, 162
model-based sonification, 173
monitorial citizenship, 48
Morena, Nina, 148
Moss, Thomas, 34
Mother Cyborg (aka Diana Nucera), 148
multi-sensory data experiences
  broadening engagement and access via, 194–198
  LAPD projects as, 210
  multiple ways of knowing via, 169–171, 177, 196
  opening doors via, 14, 166, 210, 213
  Ripple Effect as, 14, 167*f*, 167–170, 173, 213
  smellification in, 180–187, 197, 213
  sound/sonification in, 167–168, 172–180, 186, 197, 213
  taste in, 187–194, 197, 213
Muñoz Aristizabal, Martha, 16–20
Mural Arts Philadelphia, 52–53, 56
murals. *see* data murals
Museum of Modern Art (MoMA, NY), 42, 162–163
museums. *see also* pro-social spaces
  cultural reflection by, 38–43, 39*f*
  data and, 42–43, 43*f*
  educational, 41–42
  as gathering spaces, 40–41
  impact of data from, 21, 29, 38–43, 39*f*
  LAPD interactive exhibit on housing rights by, 14–15, 199, 203–210, 205*f*, 206*f*
  layers of reading via interactive exhibits in, 13–14, 40, 135–136. *see also* interactive exhibits

multi-sensory experiences with, 185–186, 197
music, 172, 175, 176–177, 179–180
MyPosition, 97*f*, 97

National Museum of Singapore, 185–186
Naur, Peter, 27–28
Neuhoff, John G., 175–178
Neurath, Otto and Marie, 42, 43*f*, 221 n.65
news organizations. *see* journalism
Newspaper Theatre, 160–161
New York Times
  data sculpture of, 129–130
  sonification by, 179
  You Draw It series, 57–58
Nightingale, Florence, 211
normal law of errors/normal distribution, 115*f*, 115–116
numerical literacy, 99

objectivity, 38, 72
O'Brien, Ruth, 89, 90*f*
Offenhuber, Dietmar, 81, 82*f*
Okdeh, Eric, 52–53
olfactory icons, 182. *see also* smellification
olfactory interface, 182–183
olfactory overload, 186
100% City series, 162
Onuoha, Mimi, 148
Open Data Institute, 46, 129
open data movement, 8, 45–47, 148
OPEN Government Data Act (2018, US), 8
open-source investigations, 37
Oppenheimer, Frank, 41
OSpace project, 182–183
ostracon, 94, 95*f*

Palacin Siva, Victoria, 150–151
Papert, Seymour, 127, 155–156, 231 n.61, 171
parameter mapping, 173
participation. *see also* engagement; inclusion
  barriers to. *see* barriers to participation
  building processes to support, 85–87
  civil society organizations encouraging, 48, 56
  data fashion shows inviting, 134–135
  data feminism and, 13, 71–73, 212
  data for good and, 13, 61–64, 212
  data justice and, 13, 67–71, 212
  data literacy and, 13, 64–67, 65*f*, 79, 212
  data murals for, 13, 51–56, 74*f*, 75*f*, 76*f*, 78*f*, 81*f*, 82–85, 83*f*, 84*f*, 212
  data protests increasing, 17–19
  data sculptures for, 114–116, 127–129

  data theatre inviting, 160–163, 207–208
  democracy dependence on, 43–46, 196
  as design goal, 55–56
  focus on, 13, 51, 207–208, 212
  impact of data to increase, 12–13, 29, 45
  LAPD projects encouraging, 207–208
  layers of reading and, 136
  mirrors increasing, 97–98
  multi-sensory data experiences increasing, 168–169
  participatory data graphs encouraging, 57–58
  participatory theatre inviting, 160–163, 207–208
  participatory urban planning inviting, 58–61, 59*f*, 114
  popular data to broaden, 7, 11–13, 207–208, 212. *see also* popular data
  pro-social spaces inspiring, 56–61, 67
Peck Evan, 146
Perovich, Laura, 150–151
physical infographics, 157
physicalization, 111
Piaget, Jean, 155–156
Pizz'age, 191*f*, 191–192
playback theatre, 161
play/playfulness
  data literacy through, 109
  data sculptures and, 114–115
  data storytelling via, 3–6, 129
  data theatre and, 150–151, 162–163
  LAPD project including, 199–203, 205–210. *see also* mini-golf courses
  mirrors of data as, 97
  multi-sensory data experiences including, 195–198
  popular data focus on, 10, 199, 208
plays, theatrical. *see* data theatre
pluralism, 44
  epistemological, 171
police, protests against, 109–110
political equality, 44
popular data
  as arts-based approach, 10. *see also* arts-based approaches
  building mirrors in, 13, 88, 208–209, 212. *see also* mirrors
  centering impact in, 12–13, 16, 206–207, 211–213. *see also* impact of data
  creating layers of reading in, 13–14, 131, 209–210, 212–213. *see also* layers of reading
  development of, 7, 9–10
  empowerment via, 10–11, 205–206, 210–215
  focus on participation in, 13, 51, 207–208, 212. *see also* participation

246 INDEX

popular data (*Continued*)
  goals of, 11–12
  imagining applications for, 213–215
  key principles of, 211–213
  opening multiple doors in, 14, 166, 210, 213.
    see also multi-sensory data experiences
  practicing, 14–15, 199
  resources informing, 7–8
popular education, 9–10, 101, 150, 160
popular sovereignty, 44
Posters for the People, 52–53
power. *see also* empowerment
  Big Data as tool for, 28
  civil society organizations and, 48–49
  data accessibility connections to, 9, 11,
    213–214
  data feminism on, 71–72
  data for good and, 63–64
  data justice to rectify imbalances of, 69–71
  data literacy as path to, 100–102
  data sculptures shifting, 115
  definition of, 28
  democracy and, 44–45
  impact of data for, 21–22, 28, 50
  of journalism, 38
  mirrors of data shifting, 13, 91
  participation balancing, 60
  popular data to rebalance, 215
practical knowledge, 170–171, 196, 210
presentational knowledge, 170–171, 177,
    195–196, 210
Priestley, Joseph, 211
Project Harvest, 166–167
propositional knowledge, 170–171, 195–196,
    210
pro-social spaces. *see also* civil society
    organizations; government; journalism;
    libraries; museums
  data literacy support in, 101, 107–109
  data sculptures in, 130
  definition of, 20
  informal learning in, 22–23
  learning about impact of data from, 12–13,
    20–23, 29–49
  mirrors of data reflecting, 97
  multi-sensory data experiences in, 178–179,
    194, 197
  participation inspirations from, 56–61, 67
protests. *see* data protests
Prudential, 159–160, 232 n.80
Putting Green, 201

qualitative data, 140
quantitative data, 140

Quick, Miriam, 179
quipu, 111–113, 112*f*

racist data
  data justice to address, 68, 71
  data protests and, 109–110
  data theatre exploring, 151–153
  impact of data to support, 21–22
  journalists reporting on, 34–35, 38
  museum exhibit use of, 221 n.65
  redlining reflecting, 85–86, 86*f*
  as windows for observation, 13, 88–98, 212
  women's clothing sizes and, 13, 88, 89–91,
    90*f*, 212
Radke, Heather, 89
Ramírez-Andreotta, Mónica, 166–169
Rand McNally, 49–50
Rastreadoras del Fuerte, Las/Trackers of Fuerte,
    The, 16–17
reading, layers of. *see* layers of reading
redlining, 85–86, 86*f*
remembrance, data protests for. *see* data protests
Research-based Theatre (RbT), 152
retirement planning, 158–160, 232 n.79, n.80
REVEAL-IT!, 96*f*, 96–97
Ribbons Experiment commercial, 159–160
Rimini Protokoll, 162
Ripple Effect, 14, 167*f*, 167–170, 173, 213
Ritter, Helge, 173
Roepstorff, Kristine, 30
Rojas, James, 114
Rosenberg, Daniel, 26
RStudio, 91, 144
Rubin, Ben, 162–163

Sayeed, Shahbaaz, 134
Scandinavian aesthetic, 139–140, 145–146
scent-based data. *see* smellification
scientific literacy, 99, 107
sculptures. *see* data sculptures
Semiology of Graphics (Bertin), 141–142, 142*f*
Shelton, William, 89, 90*f*
Simon, Nina, 41
Sinding, C., 152
SIS (Social Impact Studios), 52–53, 56
1,659 sculpture, 123–125, 124*f*
Skid Row housing rights, 14–15, 199–210
Skid Row Now and 2040 plan, 203–204
slavery
  colonialism and, 63
  data sculptures on, 125–127, 126*f*
  history of data on, 9
  journalism and legacy of, 33–34
  literacy and, 100–101

INDEX 247

Slave Trade Act (1807), 127
smellification
   arts and museum applications of, 184–186
   challenges of, 187
   community and, 186–187
   effects of, 183
   multi-sensory data experiences
     with, 180–187, 197, 213
   research informing, 182–184
   taste and, 188
   taxonomy for, 181
   touch and, 180–181
smell maps, 185f, 185
"Smell of Data" project, 184–185
Smethurst, Gamaliel, 157
Snyder-Young, Dani, 152
social chemo signaling, 180–181
social-ecological model of health, 207f
social good, data for. *see* data for good
Social Impact Studios (SIS), 52–53, 56
social justice, 70–71, 178–179
Social Justice Centres Working Group, 82–83,
   127–128
Socrates, 100
soldiers' deaths, data mural on, 75, 76f
somatic phenomenon/somatic
   memory, 155–156
sonification
   artistic, 175–176. *see also* music
   barriers to, 175
   community and, 179–180
   digital tools for, 177–179, 178f
   empirical, 175–176
   impact of, 30
   journalism using, 172–173, 175, 179
   multi-sensory data experiences
     with, 167–168, 172–180, 186, 197, 213
   music and, 172, 175, 176–177, 179–180
   scientists as resource on, 174–176
   smellification to reinforce, 186
   taxonomy for, 173
Sonification Archive, 178–179
Sort of Joy, A, 162–163
sound, data as, 30, 167–168, 172–180, 186, 197,
   213. *see also* sonification
statistical literacy, 99, 107
Staubmarke/Dust Zone (Offenhuber), 81, 82f
Stefaner, Mortiz, 191, 193
Strličlay, Matija, 185–186
Sustainable Development Goals, 92–93, 100
Svelte, 37
Swift, 80–81, 81f
symbolic data sculptures, 117–119
Symbolic Olfactory Display, 182

tableau, in data theatre, 152–153
Tableau Desktop, 141–144
tables, history of data in, 24–25, 25f. *see also*
   charts and graphs
Tactical Technology Collective, 42
Takahaski, Alexis, 148
Talmadge, Victor, 152
Tan, Froukje, 184–185
Tanzania Bora Initiative (TBI), 133, 229 n.7
taste
   community and, 193–194
   culinary design and, 191–193
   cultural context and, 187–193
   edibilization and, 189–190, 192–194, 197,
     213
   multi-sensory data experiences
     with, 187–194, 197, 213
   research findings on, 189–191
   smell and, 188
Taste of Data series, 192–193
*Te Nombré en el Silencio/I Named You in the
   Silence* (film), 16–17, 19–20, 22–23
Textbook Count project, 47–49, 56
theatre. *see* data theatre
third space, 41
Thorp, Jer, 162–163, 214
3D charts, 118–120, 119f, 224 n.47
Tidy Street Project, 74–75, 75f, 96–97
timelapse, 173
touch, 111–114, 112f, 112f, 113f, 180–181
Touwa, Winifrida, 134
Tufte, Edward/Tuftean Consensus, 137–141,
   145–146
Tukey, John, 27–28, 137
Turkle, Sherry, 171
2D representations, 141–144, 211, 212–213. *see
   also* charts and graphs
TwoTone software, 177–178, 178f

UFO (unidentified flying object)
   sightings, 104–106, 105f
Uhuru B, 80–81, 81f
urban planning
   data theatre on, 152–154, 153f
   LAPD popular data on, 14–15, 199–210
   participatory, 58–61, 59f, 114

Valkanova, Anna, 96–97
Vega, 141–142
verbatim theatre, 161
vertically integrated policy monitoring, 47
viewpoints approach, 152–153
viScent, 182–183

248 INDEX

visual encodings, 136, 141–142, 143f, 144–146
voting, 71–72, 97f, 97, 127–128

Wachata Crew, 84
Walker, Raechel, 101–102
Want, Sophie, 148
Washington, Anne L., 61–62
Wason, Oliver, 152
wealth, 9, 194–196, 195f
Weil, Stephen, 41
Wells, Ida B., 34–37
What If We...?, 176–177
Wijnsma, Leanne, 184–185
Wilberforce, William, 125–127
windows, data as, 13, 88–95, 98, 212
women
  clothing sizes for, 13, 88, 89–91, 90f, 212
  computer science lacking, 171
  data fashion show on violence toward, 13–14,
    131–136, 132f, 133f
  data murals inspiring, 84
  data protests for missing loved ones of, 19f,
    22–23, 211–212

data theatre on gender issues for, 162–163
data visualization omitting, 140–141
education/literacy denial for, 100–101
feminist movements of, 71–72, 224 n.43. *see
  also* data feminism
Woo, Rosten, 201
work. *see* employment
workshops
  Data Cuisine, 191–192
  data literacy in, 108–109
  data sculptures as tools for, 114–115
  data storytelling, 5
  data theatre, 151, 154, 162, 200
  libraries hosting, 29–30, 33
World Food Programme, 26
WTFCSV, 104–106, 105f

Yilmaz, Alyse, 191f, 191–192
Youden, W. J., 115–116

Zellner, Moira, 152–153
zines, 147–150, 149f